CODE

SIGN, STORAGE, TRANSMISSION
EDITED BY JONATHAN STERNE
AND LISA GITELMAN

Bernard Dionysius
Geoghegan

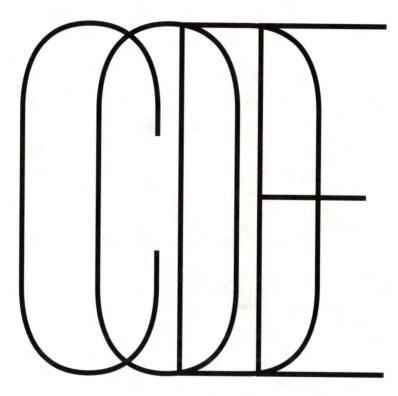

From
Information
Theory to
French Theory

DUKE UNIVERSITY PRESS

DURHAM AND LONDON 2023

© 2023 DUKE UNIVERSITY PRESS
Printed in the United States of America on acid-free paper ∞
Project editor: Bird Williams
Designed by Matthew Tauch
Typeset in Garamond Premier Pro by Westchester Publishing Services

Library of Congress Cataloging-in-Publication Data
Names: Geoghegan, Bernard Dionysius, 1970- author.
Title: Code : from information theory to French theory / Bernard Dionysius
Geoghegan.
Other titles: Sign, storage, transmission.
Description: Durham : Duke University Press, 2023. | Series: Sign, storage,
transmission | Includes bibliographical references and index.
Identifiers: LCCN 2022020276 (print)
LCCN 2022020277 (ebook)
ISBN 9781478016366 (hardcover)
ISBN 9781478019008 (paperback)
ISBN 9781478023630 (ebook)
Subjects: LCSH: Digital humanities—Political aspects. | Digital humanities—
Social aspects. | Digital media—Social aspects. | Digital media—Political
aspects. | Humanities—Methodology. | Cybernetics—Political aspects. |
Cybernetics—Social aspects. | Information society. | Technology and
civilization. | BISAC: SCIENCE / History | SOCIAL SCIENCE / Media Studies
Classification: LCC AZ105.G464 2023 (print)
LCC AZ 105 (ebook)
DDC 303.48/33—dc23/eng/20220715
LC record available at https://lccn.loc.gov/2022020276
LC ebook record available at https://lccn.loc.gov/2022020277

Cover art: Vera Molnár, *Untitled*, 1972. Computer drawing,
30 × 30 cm. Courtesy of The Mayor Gallery, London.

This book is dedicated to my very first reader and editor, Rhoda Geoghegan.

All the repetition and incarnation of the sanitized term *information*, with its cleansing cybernetic properties, cannot wash away or obliterate the fundamentally dirty, semiotic, semantic, discursive character of the media in their cultural dimensions.

Stuart Hall, "Ideology and Communication Theory," 1989

Contents

Acknowledgments

Every time I try to put to page the people to whom this book is indebted, I realize anew how many omissions remain. Many friends read chapters and discussed ideas late into the night and through correspondences that unfolded over years. Advisors and professors shared their insight and encouragement; anonymous reviewers and funding panels kept this work going across borders, languages, and programs of study. The most offhand remarks from a colleague sometimes led to far-reaching revisions undertaken in Evanston, IL; Chicago, IL; Cambridge, MA; New Haven, CT; Weimar; Paris; Berlin; Dublin; London; and Gothenburg. It is difficult to gather all these inputs together and give them the acknowledgment they deserve.

Special thanks are due to my PhD advisors, Samuel Weber, Bernhard Siegert, Jennifer S. Light, and Ken Alder, who offered formative intellectual guidance, not least of all on the dissertation that first outlined some of the ideas presented here. Great thanks are also due to my other professors and advisors for their tremendous patience as I sought to correlate ideas from various seminars and programs. Among the many teachers who had a hand in the final shape of this work, Lynn Spigel, Justine Cassell, Mark B. N. Hansen, W. J. T. Mitchell, Jeffrey Sconce, James Schwoch, Mats Fridlund, Jamie Cohen-Cole, David Mindell, Bruno Latour, and the late Bernard Stiegler are due particular thanks. Thanks also to friends Etienne Benson, Bridget Hanna, Michael Graziano, Kristina Striegnitz, Dan Knox, Marcell Mars, Elaine Yuan, Dubravka Sekulic, Tomislav Medak, Timm Ebner, Christoph Rosol, Julia Heunemann, Christina Vagt, Julia Ng, Steven Tester, and Florian Sprenger. I particularly thank John Tresch for his advice on making cuts to the original, much lengthier manuscript. I am also grateful to Karen Darling (University of Chicago Press) and Katie Helke (MIT Press) for their enthusiastic engagement with earlier versions of this manuscript. My editor at Duke University Press, Courtney Berger, offered sage and delicate guidance from acquisition to completion. I also thank her colleagues Sandra Korn, Lisl Hampton, Bird Williams, and series editors Jonathan Sterne and Lisa Gitelman.

Colleagues at the American University of Paris, Humboldt University of Berlin, Coventry University, King's College London, and Yale University offered rewarding intellectual contexts for continuing this work. I offer especial appreciation to Francesco Casetti (who read and commented on multiple drafts of this manuscript), Paul North, Gary Tomlinson, Paul Kockelman, Mal Ahern, Christian Kassung, Miriam de Rosa, John Durham Peters, Lisa Messeri, Joanna Radin, Virginia "Ginnie" Crisp, Jonathan Gray, Tobias Blanke, Waddick Doyle, Jayson Harsin, Mark Hayward, Stuart Dunn, Mercedes Bunz, Daniel Nemenyi, Lochlann Jain, and two of my key London companions in media analysis, Erika Balsom and Seb Franklin. Fellowships at the IKKM (Weimar), the Whitney Humanities Center (New Haven), and CONNECT (Dublin) provided vital financial support and intellectual companionship, including that of Christiane Voss, Birgit Schneider, Antonio Somaini, Lorenz Engell, Dennis McNulty, and Linda Doyle. From the invisible college of colleagues scattered around the world whose enthusiasm did much to sustain this work, I particularly thank Fred Turner (who generously read the whole manuscript), Geof Bowker, Mats Fridlund, Colin Burke, Mara Mills, Ben Peters, Erica Robles-Anderson, Seth Watter, the late Christopher Johnson, Theodora Vardouli, Olga Toulomi, Michael Buckland, Erhard Schüttpelz, Hans Ulrich Gumbrecht, Stefanos Geroulanos, Leif Weatherby, Noam Elcott, Alexander Galloway, Peter Sachs Collopy, Paul Edwards, Eric C. H. de Bruyn, Ted Underwood, Nathaniel Zetter, Lev Manovich, Marc Kohlbry, Jeffrey Mathias, Ron Kline, N. Katherine Hayles, and Henning Engelke. Colleagues at the Haus der Kulturen der Welt, particularly Katrin Klingan and Desiree Foerster, also provided a key setting for my work. For editorial help, and much more, I thank Paul Michael Kurtz, Alison Hugill, Laura Keeler, and Sean DiLeonardi. This research was further supported by the Jacob K. Javits fellowship program of the United States Department of Education, the Deutsche Forschungsgemeinschaft, the Melvin Kranzberg Dissertation Fellowship of the Society for the History of Technology, and the graduate schools of Mediale Historiographien and Northwestern University.

Many archives and archivists made this work possible, including Carmen Hendershot at the New School Archives and Special Collections, Debbie Douglas and Nora Murphy at the MIT Museum and MIT Libraries Distinctive Collections, Tom Rosenbaum and Bethany Antos at the Rockefeller Archive Center, and Wendel Ray at the Don D. Jackson Archive.

Nora Bateson, Phillip Guddemi, and the Bateson Idea Group provided encouragement and permission to access the Bateson papers. John Cook of Gawker furnished me with the FBI files of Claude Lévi-Strauss, which he had secured through a Freedom of Information Act request. Additional archival materials were provided by the Library of Congress, the CIA, the FBI, and the New York Public Library. I thank the communication engineers and linguists who gave me their recollections, including Noam Chomsky, the late Robert Fano, Bob Gallagher, the late David Hagelbarger, the late Morris Halle, James Massey, the late John McCarthy, and Sergio Verdú.

Editors and reviewers at *Grey Room*, the *History of the Human Sciences*, *Critical Inquiry*, and *History of Anthropology Review* provided illuminating feedback on article drafts that found their way into this book in revised form. I thank these publications for permission to republish that work. For the most part, those texts have been revised, chopped up, and rearranged such that they don't clearly map onto single chapters. These earlier drafts include "Architectures of Information: A Comparison of Wiener's and Shannon's Theories of Information," in *Computer Architectures: Constructing the Common Ground*, edited by Theodora Vardouli and Olga Touloumi (London: Routledge, 2019), 135–59; "Textocracy, or, the Cybernetic Logic of French Theory," in *History of the Human Sciences* 33, no. 1 (2020): 52–79 (which loosely correlates with chapter 5); "The Family as Machine: Film, Infrastructure, and Cybernetic Kinship in Suburban America," *Grey Room*, no. 66 (Winter 2017): 70–101 (which loosely correlates with chapter 2); "From Information Theory to French Theory: Jakobson, Lévi-Strauss, and the Cybernetic Apparatus," *Critical Inquiry* 38, no. 1 (2011): 96–126; "Nine Pails of Ashes: Social Networks, Genocide, and the Structuralists' Database of Language," *History of Anthropology Review*, August 2, 2021; and "La cybernétique 'américaine' au sein du structuralisme 'français,'" *La revue d'anthropologie des connaissances* 6, no. 3 (2012): 335–51.

I worked out portions of this book in talks and conferences, including Pierre Mounier-Kuhn's Seminaire sur l'histoire de l'informatique at Paris-Sorbonne University; Wolfgang Ernst's Medienwissenschaft Kolloquium at Humboldt University; the New Media Workshop at the University of Chicago; the Science-Technology-Culture Research Group at Nottingham University; the HASTS and Comparative Media Studies Colloquia at MIT; the Claude Shannon und die Medien Tagung at the Museum für Kommunikation Berlin; Philippe Fontaine's Postwar Social Science workshop at the École normale supérieure de Cachan; Film and Media Studies at the University of California, Berkeley; Stanford University's media studies workshop; the Franke Institute for the Humanities at the University of Chicago; the University of Cambridge Centre for Research in the Arts, Social Sciences and Humanities; the interdisciplinary media workshop at Cornell University; the Center for 17th- and 18th-Century Studies at UCLA; and annual conferences for the Society for the History of Technology; the Society for Literature, Science, and the Arts; and the Society for Cinema and Media Studies. I am greatly indebted to the participants, hosts, and organizers whose feedback and support shaped so much of this book.

Lisa Åkervall and Lucian Liam sustained me through this work. My late father, Patrick Terrence Geoghegan, offered inspiration for its investigation.

Introduction

Codification

Three grim human enclosures—laboratories, really—provided models for efforts to theorize society in digital terms. The *colony* acted as a human enclosure for theorizing data-driven cultural adjustments on a national and even planetary scale. As anthropologist Margaret Mead wrote in 1953, computing, communication engineering, and cybernetics focused ethnographic attention on communicative flows permeating "the entire network of human relationships."[1] The *asylum* provided an enclosure for documenting how codes and communication shaped human thought. In 1958, anthropologist Gregory Bateson declared, "Today, data from a New Guinea tribe and the superficially very different data of psychiatry can be approached in terms of a single epistemology" revealed by cybernetics and communication theory.[2] The *camp*, particularly the death camp, acted as a warning against states wielding unchecked technical power over life. In his 1948 book *Cybernetics*, MIT mathematician Norbert Wiener explained that the technical science he and his colleagues developed commanded "great possibilities for good and for evil. We can only hand it over into the world that exists about us, and this is the world of [the German death camp] Belsen and Hiroshima."[3] His colleague, the neurophysiologist Warren McCulloch, went one step further, arguing that it fell to scientists to construct a social order with sufficient cybernetic feedback to prevent a next, perhaps final and totalizing, world holocaust.[4]

The colony, the asylum, and the camp embodied the terrific threats of technocratic power run amok, of the state's data-driven exercises to control human conduct. They also, however, exercised allure as delimited social milieus in which supposedly controlled borders and simplified living conditions permitted the study of human communication in its elementary forms. Perhaps they even heralded the conditions of some future technological world in which all life on Earth might be subject to computer-regulated networks of communication and control. If so, was it not the responsibility of socially conscious intellectuals to look at these worlds head-on and uncover the codes and relays by which they achieved felicity or destruction?

With these examples in mind, and the fate of the world at stake, Mead, Bateson, Wiener, McCulloch, and their colleagues, aided by a chain of philanthropies funded by some of the world's wealthiest industrial magnates, rallied anthropologists, psychologists, linguists, philosophers, semiologists, physicists, physicians, and literary critics in a new epistemic machinery: the cybernetic apparatus. Refugee intellectuals, some of whom had fled fascist genocides and authoritarian pogroms or witnessed decimation wrought on indigenous persons in the Global South, flocked to the techno-political call of cybernetics. By the early 1950s, enthusiasts of cybernetics and information theory comprised a transatlantic axis of well-connected researchers winning sought-after appointments at elite universities, museums, and hospitals. By the end of the 1950s, their cybernetic jargon of *code, communication, computing, feedback*, and *control* would form part of the chatter in scientific and policy discussions across North America and Europe. They recast thought with troubling links to biological racism and eugenics in cultural systems and codes. Their turn toward communicative structures embodied an effort to develop more enlightened analytics for the force wielded by science and the state. In the course of the 1960s, a motley crew of intellectuals allied with movements such as literary semiology and structural psychoanalysis generalized this work to fields such as poetics and dementia. In the decades to follow, even as the cybernetic movement hollowed out, the promise of a theoretically rigorous approach to communicative codes endured, an unacknowledged inheritance surfacing in fields as varied as cognitive behavioral therapy in California and poststructural Marxism in Paris.

Code: From Information Theory to French Theory shows how efforts to formulate expert and technical responses to grave political crises drove the reciprocal transformation of the natural and human sciences in the twentieth century. In particular, it traces the dark industrial and colonial crises that bound adherents—including scientific philanthropies, social scientists, philosophers and literary critics, natural scientists, and engineers—in a common epistemic cause that celebrated digital research as a basis for confronting political violence. Rejecting familiar narratives of cybernetics and computing as the offspring of World War II engineering, I argue that liberal technocrats' 1930s dreams of eliminating aberrancy through noncoercive communicative techniques oriented the informatic programs of World War II and the Cold War. That dream and its paradoxical outcomes started in the rise of Progressive Era philanthropies committed to supporting communication research, traveled through media-driven studies of colonies and mental patients from the 1930s through the 1950s, drove the rise of laboratories in Cold War linguistics and anthropology, and finally roosted in the semiotic adventures of 1960s Paris intellectuals. Across these endeavors, an organizing concept of *code*—indexed to computing but derived from technocratic social science—lent proponents a powerful trope for reinterpreting the global subjects of the human sciences.

This book recovers the political impetus behind these long-running efforts to render diverse phenomena—culture, literature, illness, speech, kinship—as forms of code. My analysis encompasses the political aspirations that enveloped early efforts in information theory and electronic computing as well as the rise of so-called French theory in 1960s France. These projects, along with a wider family of programs in colonial anthropology, mental health, linguistics, literary studies, and semiotics, form diverse facets of a single political project. The roots of the project lie in Progressive Era technocracy and its agenda to transform social strife into matters for technical problem-solving.

My analysis covers the period from 1930 to 1970, occasionally touching on prehistories and afterlives outside that frame. Chapter 1 looks at robber baron philanthropies' efforts to transform the humanities and social sciences into a single field, the human sciences, oriented toward communication. It reconstructs how fields as varied as cybernetics, criminology, eugenics, and ethnography emerged as elements of an interlinked program of technocratic reform, aided by research administrators such as Warren Weaver of the Rockefeller Foundation. Chapter 2 examines how the 1930s colonial ethnography of Mead and Bateson developed the outlines of cybernetic analysis in photographic and filmic studies of Balinese tribes. Their analyses subsequently served as the basis for Rockefeller- and Macy-funded efforts to establish cybernetic approaches to the human sciences.

Chapters 3 through 5 focus on intellectual relays linking Moscow, Prague, Paris, New York, Cambridge (Massachusetts), São Paulo, and tribal territories of the Brazilian interior. Chapter 3 considers refugee structural linguists and ethnographers in America, most notably Russian linguist Roman Jakobson, whose flight from Soviet pogroms and Nazi genocides led him to embrace communication theory as a new foundation for the human sciences. Chapter 4 focuses on how French anthropologist Claude Lévi-Strauss's effort to develop a scientific and administrative response to cultural destruction wrought by colonialism, World War II, and postwar globalization shaped his ambivalent call for a cybernetic approach to the study of man. Chapter 5 considers French modernization from the 1950s to the 1970s (including the rise of laboratories, centers, and seminars as a centerpiece of French intellectual life) and how it figured in key French intellectuals' reception of cybernetics, information theory, and communication theory. Against the backdrop of decolonization, modernization, and ascendant technocracy, theorists including psychoanalysts Jacques Lacan and Luce Irigaray, literary semiologist Roland Barthes, and philosopher Michel Foucault thematized the material and political operations responsible for cultural codes. Even as their work critiqued technocracy, however, their analyses entrenched the technocratic agenda behind US philanthropies' support for communication theory. The book's conclusion argues that these histories form a common inheritance for latter-day digital humanities, cultural analytics, critical theory, and social media, and that we have much to learn from the political and ethical dilemmas that animated the cybernetic apparatus.

As a historical and theoretical study, *Code* brings information theory and French theory down to earth, situating their proponents within a landscape of cultural and political crises that drove their efforts to master the unruliness of human communication with technical expertise. This book seeks to dislodge academic theories from an impulse toward universality that inflects much scientific inquiry, and instead situates these efforts as responses to determinate political crises that were neither universal nor general.[5] In this respect, *Code* is a somewhat classical exercise of critical theory in the sense of reconstructing the historical problems and political interests that organize a system of knowledge and that lend that system an appearance of disinterested generality (particularly through its complementarity to a larger system of economic production).[6] All theories, of course, involve some degree of abstraction as a preliminary step in relating diverse specific instances to a more general condition. An ongoing task of theory (and history) is to develop and intermittently reassess such abstractions according to changing circumstances.[7] In that respect, this book, if successful, is not only an assessment of, but also a contribution to, theoretical reflection. Yet some adherents would deprive theory of this worldliness, preferring to appoint this or that theoretical system as total and complete. The deadening result of such methodological dogmatism is to render theories primarily of theoretical interest, depriving them of contact with the situated struggle that allows them to know and intervene in the world. When that happens not only theory, but theorists, too, become remote and unworldly. Against such hermeticism *Code* insists on theory's place in uneven and changing fields of scientific and political struggle, in which it plays the part of participant as well as observer.

Knowledge Lost in Information

What is the reality of a sign? In a world marked by the proliferation of technical communication, at what point must signals be considered not representations but full-fledged objects for scientific and political calculation? These questions preoccupied the natural and human scientists of the twentieth century, motivating them to turn to one another for provisional answers. Technical advances in telephony, computing, propaganda, and the mathematics of signal transmission renewed the urgency of such questions. Likewise, the inevitability of cultural confrontation at a distance—exemplified by colonial command and control across oceans or dueling fascist and democratic propaganda—outlined a problem intractable to traditionally conceived politics. Confronting these challenges drove mathematicians, anthropologists, psychotherapists, linguists, and semiologists to reconsider the sciences writ large. As scientists formed interdisciplinary working groups to address these questions, they aligned their investigations with external funding—and, in turn, priorities—made available by private foundations, their officers and executives, and myriad strategic initiatives cutting across the scientific, political, and military spheres. "Theory" provided one means of formulating research programs that could navigate the institutional thickets of scientific funding circa 1930–1970. Read as transhistorical statements, notions of communication propounded by the likes of Mead, Bateson, Jakobson, or Barthes might appear bizarre. Read as interventions in the knowledge machineries of their day, they take on the profile of canny political performances that affirmed the risk of intellectual engagement in their contemporary ethical concerns.

The conceptual itinerary running from information theory to French theory is partly a history of how technocracy shaped the language and professional norms of responding, as an academic or intellectual, to political urgency and in particular tracks how informatics offered a tempting framework for ordering those norms, especially when the underlying intellectual models stemmed from politically motivated research into nonnormative populations. Interest in code, a theme that recurs across the chapters in this book, reflected one ideal of scientific precision at a time of technocracy and ascendant information sciences. Particularly after World War II, progressive researchers embraced communication theory as a resource for securing a place for the human sciences among the then emergent technical sciences. This project had contradictory outcomes with peculiar resonances across disciplines and political cultures. For example, as I recount in chapter 5, in 1967 Barthes launched a seminar in Paris at the École des hautes études en sciences sociales dedicated to the close reading of codes in Honoré de Balzac's novella *Sarrasine*, a work rife with the contradictions of class, gender, and narrative. For Barthes, word-by-word reading revealed the pervasive codification of language as well as the cacography, countercommunication, and noise confounding its signals.[8] Sociologist Daniel Bell of Columbia University in New York wrote that same year, "What has now become decisive for society is the new centrality of theoretical knowledge, the primacy of theory over empiricism, and the codification of knowledge into abstract systems of symbols that can be translated into many different and varied circumstances."[9] On two sides of the Atlantic, analysts of diverse stripes embraced theory as a way of interrogating how far analysis might go in defining order and disorder through a theory of cybernetic codes.

Scientists' cybernetic efforts to reconstruct society in terms of totalizing codes outpaced existing communication technologies. As of 1970, no data system adequate to the task existed. If it had, no liberal democracy would have tolerated its deployment. The closest thing to a real-world implementation of such a system—the US military's shoddy surveillance systems in the jungles of Vietnam—proved no match for popular resistance.[10] This restricted the actual deployment of cybernetic methods to specialized domains such as the colony, the clinic, the asylum, homes of the mentally ill, so-called ghettoes, and, as it turned out, literary seminars. This lent a speculative quality to midcentury communication research due to its proponents' penchant for generalized claims based on data from the colonized, the institutionalized, and the fictionalized. In the conclusion of this book, I ask how those constraints remain operative in an era of cultural analytics and pervasive social networking.

Cybernetics' Technocratic Backstories

A major goal of the book is to recover the history of the human sciences as a test bed for the rise of the communication sciences, as realized in the experimental systems of imperialism, colonialism, and industrial capitalism. To adapt a formulation from feminist historian of science Donna Haraway, the informatics of domination began not in informatics but in domination.[11] Progressive Era political crises—which is to say, liberal expert programs to save industrial democracy from racial, economic, and other inequities threatening the integrity of the entire social order—drove a technicization of the human that swept across policy and science in the early decades of the twentieth century. These crises shaped the meanings attributed to communication and culture across the humanities and social sciences. Indigenous myths, French literature, Russian poetry, mental illness in the American suburbs, ethnographic film and photography, and press photographs were among the exemplary objects used for working out the roles of information, code, and communication in the modern world.

Acknowledging the histories of cybernetics and the human sciences demands a reconsideration of what kinds of violence and innovation count in histories of computing. For too long, media theory and the history of science have taken armed conflict between nation-states as the mother of media invention. Groundbreaking works by historians, including Peter Galison and Paul Edwards, posited wars between Western states as the engine for digital technologies and epistemologies.[12] The technocratic roots of cybernetics in the Progressive Era and interwar years challenges that historiography. They show that political dilemmas raised by inner cities, colonies, asylums, and even zoos modeled the epistemological and technological developments of cybernetics, information theory, computing, and digital media in World War II and the Cold War.

My reorientation toward the role of interwar technocracy in cybernetic development follows a larger reconsideration set in motion by historian Jennifer S. Light's seminal study of cybernetics' role in welfare policy.[13] While evidence for the practical efficacy of cybernetics in armed conflict is limited, Light has shown that its application to urban ghettoes in the United States was broad and aggressive. In the intervening years since Light's scholarship, a booming literature has examined how Native Americans, criminals, children and other learners, colonial subjects, minority communities, people with disabilities, and animals served as key objects for theorizing the information sciences.[14] These works have definitively shown that the "enemies" and "closed worlds" decisive to cybernetics were not typically foreign nationals of the Axis but rather domestic and colonial populations within the borders of nominally democratic industrial states. These populations' precedence in the rise of cybernetics shows how colonies, asylums, and suburban enclaves figure constructively in larger histories of systems theory, game theory, computational methods, and data-driven analytics.

This enlarged context for computing history not only supplements an inadequate history that has consistently privileged engineers and their machines as the subjects of historical narrative, but it also takes steps toward a global history of information and computing that includes the suffering, strife, and participation of persons deemed less than full citizens or subjects by the state. To incorporate these sites and subjects, including their management by the human sciences, is to counter an insistent analytic that has attributed minoritized media practices to "culture" while reserving the categories of "technology" and "invention" for the activities of scientists backed by industrial laboratories and state sponsors.[15] Much work remains to be done for a historiography of technology that also grasps the diverse agencies and contributions of minoritized subjects. The present history concentrates on a small and privileged network of highly educated scientists (e.g., Mead, Bateson, Irigaray) who observed other communities put under surveillance by social science and the state. In a handful of instances, the scientists faced the prospect of their own annihilation as minoritized persons (e.g., Jakobson, Lévi-Strauss), an experience that shaped their later work.[16] This book is a partial and initial survey of the definition of computing, communication, and informatics in the United States by political refugees and ambivalent witnesses to colonial violence. As these theorists occasionally noted, myriad scientific and historical techniques militate against incorporating the voices of people enclosed by medical, colonial, fascist and other technologies. *Code* does not overcome this problem but takes modest steps toward its fuller recognition.

Recognizing the priority of liberal social science in histories of cybernetics and computing also demands that we rethink these fields' identification with movements such as posthumanism and antihumanism. We can think here of essential writings by N. Katherine Hayles, Donna J. Haraway, Friedrich Kittler, Rosi Braidotti, and Claus Pias, to which *Code* is deeply indebted.[17] These works credit the rise of informatics with fostering new attention to codes, signs, and relays not anchored in classical liberal humanist categories like body, author, and intention. Yet, often, this claim overlooks the priority of colonial ethnographers in formulating this analytic through studies in the 1920s and 1930s of practices such as gender, performance, and storytelling in indigenous communities. This has often led to a strange historiography in which Western technological inventiveness claims credit for categories borrowed from non-Western social collectives. Exemplary is media theorist Pias's account of cybernetics as an "affront to anthropology" that sought to "demystify instead of mythologize" humanity.[18] For Pias and other proponents of a hardware-oriented media theory, cybernetics—born in the mathematics and tinkering of US and German laboratories—staged a conceptual assault on humanist thought. There is much truth to such claims. The popularization of cybernetic ontologies in fields including semiotics did support a decisive turn from categories like intention and authorship to those of signification and systems. Yet it overlooks how thinkers of cybernetic systems, among them the anthropologists Mead, Bateson, Lévi-Strauss, and Lacan, drew on indigenous cultures to formulate an account of system, influencing thinkers from Wiener to French biologist Jacques Monod in the process.

A more complete historiography of cybernetics, including its global and colonial roots, complicates claims to its supposed antihumanism. No longer a break *from* Western humanism, it rather traces fissures and contradiction internal to that humanism. Where one draws a historical or epistemic break is, itself, a deeply political and historical choice. To consider World War II and computing as a great rupture is also to mark oneself off what from what came before—to free oneself, in a sense, from its entanglements. Choices about periodization entail decisions about which social, political, and ethical configurations most weigh on us today. So, too, that assignment of the break entails assignments of agency as well as of margins and centers to the historical change it sets in motion. *Code* offers a periodization attentive to how ideas about cybernetics, information, and computing on the rise in the 1940s belong to a program of technopolitical reform that took shape across the 1920s and 1930s, in connection with projects like colonial anthropology and eugenics. This analysis—which views the cybernetic critique of humanism as part of its humanist agenda—seems faithful to the insights of many of its early proponents. If, as Lévi-Strauss claimed, the goal of informatically inspired human sciences was "not to constitute, but to dissolve man," this concerned a fulfillment of global processes at work long before digital computers offered figures for their conceptual refinement.[19] That dissolution is not unrelated to the battlefields of World War II Europe, to be sure, but it also travels through a global network of colonies and asylums that formed an early training ground for thinkers like Lévi-Strauss, Mead, Bateson, and McCulloch. In our era of pervasive digital communication, when digital methods serve epidemiology, criminology, literary studies, data visualization, and social networking, it's time to recover these histories—not to find their true and hidden origins but to start grappling with their political and cultural complexity.

Cybernetics' technocratic pedigree, cultured in the human sciences, also reorients how we think about the disembodiment often identified with cybernetics and information theory. That reorientation, in turn, shapes our understanding of pervasive datafication in the present. As a critical intervention to the preoccupation with virtual reality in 1990s web cultures, literary scholars such as Hayles and Mark B. N. Hansen linked cybernetics and information theory with ideologies of posthuman disembodiment.[20] But perhaps the key erasure was more political than technological (an argument that the fuller arc of writings by Hayles and Hansen would seem to support). Like the dreams of sublime communication that shaped the early histories of networked railroads and telegraphy, dreams of cybernetic posthumanism depended on disappearing the bodies of native persons and other subjects regarded as less than human. That drive toward informatic disembodiment belongs to a liberal ideology that promises equality through the technological suspension of geographical, linguistic, ethnic, and class differences.[21] The ideology of an information society central to projects from Facebook to informatic policing updates this old political fantasy.

Cybernetic anthropology, psychotherapy, and semiotics offered an early sketch of a larger cultural analytics ascendant in twenty-first-century academia, industry, and governance. Practitioners recast phenomena as diverse as madness, suicide, dance, and poetry as informatic "codes" suitable for expert reform. As mid-twentieth-century heads of state celebrated notions of human rights with distinguished Enlightenment pedigrees, interdisciplinary alliances of anthropologists, linguists, literary critics, psychotherapists, and philosophers met in conference halls, classrooms, museum collections, and laboratories and developed theories of the cybernetic subject. In these spaces, nascent ideologies of an information society took shape decades before the large-scale implementation of information processing. Fields such as the arts, social sciences, and humanities served as experimental laboratories for the engineering of a society of information superhighways. Far from trailing behind engineers and natural scientists, human scientists spearheaded the reconceptualization of cultural forms prerequisite to the emergence of an information society, furnishing a legitimating ideology for an era to come.

The Language of Communication Science

I would like to briefly review a few key terms found throughout *Code*, with a caveat: much of the interest of cybernetics springs from its incessant resignifications, sometimes from year to year, of familiar terms. Frequently, invocations of concepts, such as encoding, decoding, information, feedback, entropy, and system, veiled the political dimensions of social analysis, lending to social scientists the appearance of cool and dispassionate scientism.[22] The bottomless well of signification of these terms, and their ability to cross disciplinary and political borders, proved a mixed blessing for proponents. As Mead said of the cybernetics conferences, "We were impressed by the potential usefulness of a language sufficiently sophisticated to be used to solve complex human problems, and sufficiently abstract to make it possible to cross disciplinary boundaries. We thought we would go on to real interdisciplinary research, using this language as a medium. Instead, the whole thing fragmented."[23] This fragmentary nature was there at the beginning of the field, corresponding to the wide spectrum of political stakes the field attracted. Landmark works by historians Steve Heims, Ronald Kline, and Geoffrey Bowker, among others, have done much to capture the process by which cybernetics manufactured, managed, and sometimes erased political and scientific differences.[24] As such, *Code* does not seek to recover an originating moment or problem that fully defined cybernetic research. My aim is rather to reassemble what I see as key, and somewhat overlooked, social attachments that shaped these fields and their significance for our conflicted present moment.

Among the terms displaying some semantic variability throughout this book are *cybernetics*, *information theory*, and *game theory*. Their meanings were particularly unstable from around 1948 to 1964. Efforts to canonize one definition or another usually speak to specific theorists' desire to monopolize credit or authority and shape specializations in the shifting grounds of postwar universities. Those goals, reasonable within the push and pull of academic politics, obscure the epistemic webs essential to tracing common endeavors from information theory to French theory. For this reason, my treatments of these terms often emphasize the periods of formative and developing methods, such as computing at MIT in the

1930s or Lévi-Strauss's work at the Paris office of UNESCO in the 1950s. The interest of these sites is as relays: sites of decisive epistemic exchange along routes that may have collapsed over previous decades of scientific specialization. I view that shifting groundwork as central in reconstructing decisive affinities typically hidden by traditional disciplinary accounts. In other words, this book is interested in mapping the diffractions and differential specializations that bind diverse fields together—what Kline has termed "the disunity of cybernetics."[25]

With those caveats in mind, by *cybernetics* I usually refer to a science that emphasizes communication and control in humans, machines, and living systems. For Wiener, as for Shannon and mathematician John von Neumann, cybernetics variously subsumed or overlapped with information theory and game theory.[26] Cybernetics closely aligns with 1940s research into electronic computing, feedback mechanisms, and biology. The term is commonly identified with Wiener, who coined its modern usage in the late 1940s; but in the course of chapters 1 and 2 I reset its meaning within the broader associations of the Macy conferences on cybernetics that supported its conceptualization, including the media anthropology of Mead and Bateson in the 1930s. Indeed, cybernetics is capacious and at times has included works by the likes of Lévi-Strauss, Lacan, and even Barthes.

By *information theory* I generally mean Shannon's account of the theoretical limits and coding schemas for data transmission as well as its reinterpretation by mathematician Warren Weaver in his celebrated commentary to their coauthored book, *The Mathematical Theory of Communication*.[27] Yet there is also a theory of information embedded in Wiener's cybernetics, much as cybernetic concerns with feedback mark Shannon's writings. *Game theory* most directly concerns the writings of mathematician John von Neumann and economist Oskar Morgenstern on games and economic behavior. But its kernel is not their work but rather a more general effort, either mathematical or social scientific, to formalize diverse behaviors through reference to (usually) simple games. Throughout this book I also touch on remarks on games by Bateson, Lévi-Strauss, Lacan, and Shannon that fall within the ambit of 1950s and 1960s game theory.

Occasionally I invoke the terms *communication science* and *communication theory*. Although somewhat familiar as phrases since the 1950s, these fields were never subject to the same disciplinary stabilization. The collapse of MIT's longstanding efforts to establish a center for communication sciences (figure 1.1), where thinkers as diverse as linguist Noam Chomsky, information theorist Shannon, and experts in fields such as servomechanisms, biology, sociology, and psychology were to be brought together, is telling. That effort speaks to both the institutional reality of these ideas and their insistently unwieldly characters. For the purposes of this book, *communication science* refers to the far-reaching effort in the 1950s and 1960s to redefine diverse disciplines in terms of communication—a project that included communication studies, structural anthropology, and cognitive science. While the present book has little to say about molecular biology or ecology in the 1960s, there are key works in those fields that could be classified as communication science. In contrast to cybernetics and information theory, communication science was not explicitly tied to industrial engineering or data transmission. Saussurean linguistics, for example, might be understood as a field of communication science. Yet an implicit reference to the technical underpinnings of cybernetics and information theory abided whenever more theoretical efforts were made to conceive of all these diverse objects as part of a common project. Hence the term *communication theory* refers to conceptual frameworks invoked by these fields to classify their objects as communicative.

By the phrase *cybernetic apparatus* I mean the network of institutions, methods, techniques, researchers, conferences, instruments, laboratories, clinics, infrastructures, and jargon mobilized around cybernetic themes. I use the term *apparatus* to evoke two aspects of that mobilization. First,

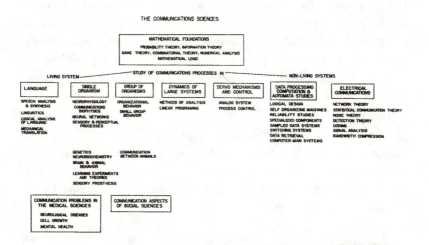

1   Draft diagram of disciplinary structure at MIT's Center for Communication Sciences, ca. 1959. Use of *communication* and *science* in singular and plural forms reflected broader conceptual instability around these key terms. Source: J. B. Wiesner Papers, box 4, folder 130, MIT Distinctive Collections, Cambridge, MA.

I have in mind the sense of an apparatus as an instrument or device. Researchers celebrated the power of new instruments such as telecommunications, computers, and electronics to reveal the organization of the world in discrete signals. They also hailed the ability of these instruments to organize transdisciplinary knowledge that transcended such instruments and techniques. The term *cybernetic apparatus* therefore refers to cybernetic instruments and techniques that permitted the mobilization of diverse objects and people within apparently cybernetic relations—for example, by revealing patterns, codes, and systems underpinning diverse elements. Bateson and the Palo Alto group used family therapy observation rooms, which included instruments such as film cameras, magnetic recording, and structured interviews, for the disclosure of recursive informatic patterns structuring social groups. These rooms and their instruments became cybernetic apparatuses. In this sense, references to the cybernetic apparatus always imply concrete instruments and techniques that translate diverse objects into cybernetic flows.

Second, with the term *cybernetic apparatus* I have in mind the strategic and political alliance of researchers around cybernetic technologies, usually supported by foundations or other interested sponsors. The apparatus always corresponds to urgent problems that mobilized researchers and institutions across disciplinary, political, and national borders by reference to the quasi-transcendental powers of cybernetic instruments. The management of colonial subjects, mental illness, the Soviet threat, the fallout of decolonization and globalization, and the rise of mass media constituted problems that mobilized an apparatus of researchers, institutions, and techniques for analysis and management. *Code* explains how the blurring of these potentially distinct types of apparatuses—instruments on one hand, a strategic convention of heterogeneous actors on the other—resulted in a new epistemic machinery. My historical reconstruction of the cybernetic apparatus casts new light on the pervasive technological and political dimensions of the communicative epistemology that spread across disciplines in the twentieth century and laid the groundwork for our digital present.

Technocracy and Theory

Throughout this book I refer to technocratic orientations of the cybernetic apparatus. Somewhat more detailed clarification is needed of my use of this term. By *technocracy* I mean a political strategy that invokes technology, data, and expertise as a means for the neutral (or nonpartisan) adjudication of political difference. Although there were a few political groups of the 1930s with *technocracy* in their names, I am using the term here to refer to a general political strategy of the period rather than to a specific group. Surveys, social sciences, laboratories, theory, and jargon often figure prominently in technocracy. Technocracy is frequently misunderstood as governance by means of technology and its related apparatus of data, bureaucrats, and nonexperts. But the core feature of technocracy is a politically motivated valorization of the technical as a supposedly nonpolitical and neutral tool of governance. It defuses social struggle by likening social conflict to mechanical failures, suitable for impartial redress by technical experts. In other words, the essence of technocracy is not the technical as such but rather a political rhetoric of the technical. Technocracy obfuscates the political through appeals to technology, techniques, procedures, data, technical precision, and so forth as tools of neutral governance.

As historian Dorothy Ross argues, the rise of the social sciences in the United States owes much to middle- and upper-class professional researchers who interpreted social and political conflict that was industrial in origin in depoliticized terms of adjustment and adaptation.[28] Technocracy appealed to liberal reformers and even to industrial magnates, who sometimes funded scientific philanthropies, universities, and similar expert institutions of technocratic reform. In a society that was plagued by capitalist inequality but that had an official commitment to equality, freedoms of belief and movement, and the protection of private property, technocratic social science seemed to promise noncoercive means for adjudicating these interests. A 1943 essay coauthored by Harold A. Lasswell, a major figure in interwar communication research and postwar cybernetic social science, praised the increasingly "technical and exhaustive" techniques of the "modern specialists on the human sciences" for furnishing knowledge that could be "applied and retested in the selecting of personnel in business, government, army, and other social structures."[29] By means of the "laboratory of the psychologist, the field expedition of the ethnologist and the clinic of the physician . . . systems of incentive are explored for their efficacy in raising production and reducing disciplinary problems."[30] This kind of technocratic social science obfuscates how industrial management sets social ideals by transforming its dictates into supposedly general theoretical knowledge about the human condition. Despite that, technocracy is not as blunt a form of industrial exploitation as strike-breaking with

Pinkerton detectives. As an instrument of technocratic governance, the human sciences displace force through the mediating instruments of numbers, laboratories, peer review, and data, giving the dictates of industrial and elite management the trappings of neutral scientific adjudication.

A particular relationship prevails between theory in the human sciences and technocracy, particularly in the United States. To characterize that relationship in very rough and general terms, theory is a means of rendering the humanities more consonant with the dictates of technocracy that obfuscates political decisions through an appeal to abstract norms. As philosopher Jürgen Habermas has argued, since industrialization—itself a process dependent on a certain institutionalization and application of theoretical knowledge—science and technology have provided a central model for nonideological governance.[31] Fields from economics to politics increasingly legitimated their professionalism via appealing to theoretical models that oriented professional practice. Indeed, adhering to scientistic theory remains one of the markers setting off professions from "mere" trades.[32] These theories and their models form an ever-advancing set of theoretical points of reference by which the disciplines submit to rules of formal analysis and definition. Particularly in fields open to contingent and singular impressions of an individual (e.g., sciences of the mind, literary criticism, media studies), theory often seems to promise a set of neutral and professional standards for ordering observations.[33] The late twentieth-century rise of "theoretical schools" in the US humanities shows that the demands of professionalization render even theories thematically opposed to such standardization, such as Marxist critical theory or deconstruction, susceptible to the development of standard machine-like techniques for predictable selection, interpretation, and publication of results. This alliance between theory and professionalization is one manner in which the contemporary university accords with the demands of a technocratic era.

These remarks on theory, technocracy, and professionalization bring up a decisive point: a false dichotomy opposes *technocratic analysis*, identified with empirical and data-driven inquiry, preferring practices of *theory* or *theoretical analysis*, allegedly defined by a speculative analysis of metaphysical or critical inflection. Yet the rise to prominence of theory in the study of humans closely tracks with efforts to develop scientistic correlations between empirical observations and systematic social critique. For example, it was amid an intense emphasis on reorienting the relationships among empiricism, social science, and critical philosophy that sociologist Max Horkheimer proposed critical theory as a new manner of orienting rational social inquiry.[34] Indeed, many modern critical projects that explicitly reject scientific positivism nonetheless resound with an effort to retool philosophy for an industrial age, including an adaptation of critical inquiry to the methodological and professional models of the technocratic human sciences. Theodor Adorno's well-known work *The Authoritarian Personality* is an exemplary effort by a critical theorist to align critical theory with rigorous and empirical social science for the purposes of improved educational policy and the like. Theoretical schools' frequent recourse to institutes, external funding, and think thanks reflects a broader drive to reorder the humanities in forms and values borrowed from the social sciences. This apparatus belongs to an effort to reconstruct critique according to the professional, institutional, and political exigencies opened up by empirical social sciences.

In so-called French theory the debts to technocratic social science were particularly evident. As Lévi-Strauss wrote of linguistics, which he treated as a model for the human sciences, linguistics "is probably the only [social science] which can truly claim to be a science and which has achieved both the formulation of an empirical method and an understanding of the nature of the data submitted to its analysis."[35] It was on this basis that Jakobson's structural linguistics, Barthes's structural semiology, Lacan's structural psychoanalysis, and to some extent Louis Althusser's structural Marxism took linguistics as the model of a modern and exacting inquiry. As discussed in chapters 4 and 5, far-reaching reforms in the 1950s and 1960s that promoted data-driven research centers, teamwork, empirical methods, and a reorientation of philosophy and literature toward the "human sciences" were a precondition for the theoretical revolutions tied to psychoanalysis, semiology, and Marxism in Cold War France. That does not necessarily mean that theorists are technocrats, but it does suggest their work was being refashioned according to the conditions of a technocratic institution craving legitimacy in industrial democracies that value professionalism and technical expertise.

The technocratic advance of theory in the human sciences in France and across Europe and North America correlated with efforts to restructure singular works—literature, art, political events—into symbolic fields structured in more or less predictable manners disclosed by organized professional study. The drive in the modern research university toward regulation by field-specific jargons, professional bodies, research centers, peer reviewed journals, conferences, and increasingly complex hierarchies of supposedly meritocratic distribution reflects the same technocratic tendency. When film theorist David Rodowick writes that "from the late 1960s and throughout the 1970s, the institutionalization of cinema studies in universities in North America and Europe became identified with a certain idea of theory," the term *institutionalization* is at least as crucial as *theory*.[36] Theory not only translates words and images into a logical system of discrete elements permitting reconstruction in the scientific frameworks implied by semiotics or psychoanalysis of the period. It also becomes a key element in coordinating new inscriptions of objects and personnel into the modern research university in accordance with professional and methodological distinctions loosely modeled on the natural and social sciences. The inscriptions carry with them jargon, hierarchies, interpretive prescriptions, authoritative texts, and other standards that discipline research communities. This need for professionalism, as Ross notes, corresponds in turn with the demand that the social sciences display the rigor, impartiality, and neutrality that liberal industrial democracies require of the modern research university. Theory thus systematizes both its subject and its object, the researcher and the researched.

The Human Sciences and the Behavioral Sciences

Throughout this work I refer to the *human sciences*. In general, it designates the idea of a scientific, systematic approach to the social sciences and humanities. Geography, linguistics, anthropology, criminology, and literary semiology, when practiced with a systematizing and theoretical emphasis, fall within its ambit. Psychiatry and psychoanalysis are also frequently within its purview. Aspects of biology may occasionally be within its mandate, particularly when, for example, they bear on a humanistic field such as physical anthropology or linguistics. I use the term *human sciences* throughout this book not only because it surfaces among the subjects I study but more importantly because it designates a conceptual effort to which my subjects implicitly subscribed. At key moments in their research careers, Mead, Bateson, Wiener, Shannon, Lévi-Strauss, Barthes, and others believed fields such as cybernetics, information theory, and semiology were uncovering general principles that underpinned natural and human systems including psychology, linguistics, and the arts. This belief led to a certain skepticism toward existing disciplinary arrangements as well as to the effort, through initiatives such as the Macy Conferences on

Cybernetics, to found new fora for dialogue between the humanities and social sciences.

Perhaps more decisively, the human sciences refer to a project of systematization long sought but never achieved in an enduring form. For example, in 1963 Althusser remarked that while everyone seemed to be talking about the human sciences, when asked what they were, the same people "have trouble coming up with an answer"; he added that, in his own estimation, "it is a field in search of its own definition."[37] To be sure, there are plenty of projects and invocations of "the human sciences" as if it were a definite and existing intellectual formation. The aforementioned remarks from Lasswell exemplify the type of context in which the idea of human sciences often thrives in the United States: among experts putting knowledge of the human in the service of government and industry through improved techniques and data. This tendency, equally on display in France, led Lacan to comment in 1965 on his "lifelong repugnance for the appellation 'human sciences' . . . it strikes me as the very call of servitude."[38] Indeed, something deep within the human sciences drives many of its most celebrated thinkers to question not only the term but the very "human" revealed by the systems, techniques, data, and enclosures its researchers cultivate.

For these reasons, with the term *human sciences* I denote an ongoing effort at reform, always underway, never realized. Like the colony, the asylum, and the camp, it strives for but never achieves closure. The aspiration to contain and order difference is stymied, not least by the social and political contradictions that set its systematization in motion. *Code* describes one particularly decisive itinerary by which the project of the human sciences was done and undone in the course of the twentieth century. The additional term *behavioral sciences*, which closely overlaps with French usages of *les sciences humaines*, provides one more example of perpetual unraveling of these scientific forms as they cross borders.[39] While some semantic differences might be inventoried among the human sciences, behavioral sciences, and cybernetic social sciences, binding them together is a tendency to undermine the objects to which they refer. These scientific programs to define humans, behaviors, and social groupings tend finally to render their order of reality uncertain and inconsistent. These fields' commitments to systematization position human beings in technical and scientific flows that no longer regard man as the measure of all things. The effort to institute a science of the human, much like the effort to reach a purified vision of "reason" or "culture" by means of the colony, asylum, or camp, leads to endeavors defined chiefly by inhumanity and barbarism. This was no hypocrisy haunting the cybernetic apparatus; on the contrary, it was the horrifying paradox that its proponents chose to face head on, by reincorporating political and technical force within the sciences of humanity, if need be. Their struggle over the human sciences self-consciously coincided with the unbearable facts of genocide, pogroms, and colonization as threats intrinsic to the digital methods to which they also turned for a way out of these human catastrophes.

One

Foundations for Informatics

Technocracy, Philanthropy, and Communication Science

The "facts" of science, which we honor as being so objective and so indisputable, are . . . "the facts" because our Western scientific philosophy has developed a language, a logic, a culture, and an ethics which, all taken together, have led us to describe nature in one particular set of abstractions. **Warren Weaver, Director of Natural Sciences, Rockefeller Foundation, 1950**

In the 1930s the world's largest private patron of scientific research, the Rockefeller Foundation, initiated a far-reaching effort to transform the humanities and social sciences. The mission was to transform "armchair thinkers" in the thrall of "medieval logic and . . . metaphysical speculation" into empirical scientists guided by observation, statistics, and instrumentation.[1] Research on communication (including propaganda, mass reproduction, radio and television, and Basic English) figured centrally in this program of reform. Foundation president Raymond Fosdick explained that "there is undoubted value for scholars in . . . an exegetical commentary on the fourth book of Virgil's Aeneid . . . [but] in this mechanized age, something more than this is needed, some method by which the esthetic and spiritual meanings of human life can be interpreted over wider areas."[2] Aware that the foundation's focus on scientific techniques and media instrumentation troubled some scholars, Fosdick later asked, "Where is the line that can be sharply drawn between technology and content?"[3]

Wherever that line stood between the two world wars, in the course of the 1940s, foundation officers took steps to erase it with digitally inspired communication research. In fields including anthropology, linguistics, mathematics, biology, and psychotherapy, the Rockefeller Foundation and a host of like-minded philanthropies funded by robber barons (e.g., the Ford Foundation; the Josiah Macy Jr. Foundation; the Wenner-Gren Foundation) lavished generous funding on interdisciplinary research linked to research programs inspired by cybernetics. A dream of rigorous cultural analytics that processed diverse objects as analogous instances of information and communication drove these initiatives.[4] The Rockefeller Foundation's natural sciences director, Warren Weaver, by training a mathematician, stood at the intellectual fore of these endeavors. His 1949 reinterpretation of mathematician Claude Shannon's article "A Mathematical Theory of Communication" championed the interdisciplinary promise of communication sciences. According to Weaver, Shannon's theory was "so general that one does not need to say what kinds of symbols are being considered—whether written letters or words, or musical notes, or spoken words, or symphonic music, or pictures. . . . The relationships it reveals indiscriminately apply to all these and to other forms of communication."[5] Through writings, lectures, travels, and funding initiatives, Weaver cultivated a coalition of researchers, journalists, policymakers, and educated members of a general public committed to the idea that many, perhaps most, phenomena could be explained in terms drawn from communication engineering. For decades to follow, when anthropologists, psychotherapists, literary critics, sociologists, art historians, and other human scientists invoked information theory, they cited Shannon, but their analyses reflected a cultural analytic forged by Weaver. In recognition of his contribution, the University of Illinois Press added Weaver's commentary to its re-publication of *The Mathematical Theory of Communication* and credited authorship jointly to Shannon and Weaver.[6]

Weaver's analytic capped decades of efforts to cultivate informatic analysis in the natural sciences.[7] During the 1930s and 1940s Weaver had green-lit or supervised key projects, including engineer Vannevar Bush's computing research at the Massachusetts Institute of Technology (MIT) and mathematician Norbert Wiener's cybernetic studies with physiologist Arturo Rosenblueth. He encouraged an interest in cybernetic themes among anthropologists Claude Lévi-Strauss and Gregory Bateson and linguist Roman Jakobson; Weaver's enthusiasm likely influenced the Rockefeller Foundation's decisions to fund their research projects. He helped establish the field of computer translation and intermittently intervened to provide financial and institutional support to far-flung scholars including the British information theorist Donald M. MacKay. His informational writings even attracted interest from the likes of Austrian economist Friedrich Hayek, who once sought to translate them into German. Weaver's systematic efforts did much more than improve the instruments and techniques of the natural sciences. He realigned scientific analysis with the kind of technical analysis favored by technocratic philanthropists. In arguing that not only scientific problems but even questions of art and human freedom could be treated in computational terms, Weaver conjured visions of a vast and messy world available for expert improvement, if only its underlying dynamics could be recast in terms suitable for technical analysis.[8]

This chapter situates the rise of informatic analysis, particularly Weaver's information theory, in terms of longstanding efforts by philanthropies born of the US Progressive Era (ca. 1890–1920) to cultivate technocratic analysis across disciplines. I reconstruct how these organizations' commitments to technical problem-solving led them to support research in computation and communication with the goal of advancing social and cultural reforms. Their work aligned problems as diverse as automated computing, the unification of the sciences, criminology, eugenics, linguistic analysis, and the methodologies of human sciences in a network of epistemic affinities whereby communication and computation offered privileged methods of problem-solving. Robber baron philanthropies' enthusiasm for cybernetics, information theory, and game theory reflected the belief that science was closing in on "technologies" by which all these diverse "contents" would submit to an integrated system of technical problem-solving. In this respect, World War II advances in computing seemed to fulfill technical needs for a Progressive Era alliance struck between so-called scientific philanthropies and a burgeoning apparatus of experts in science, engineering, and the social sciences. The largely male and often patrician network of scholars, administrators, policymakers, and businessmen celebrated the power of technical ingenuity to dispel poverty while leaving broader economic and political structures intact.

This history of robber baron philanthropies suggests that welfare, rather than warfare, provided an impetus for the midcentury rise of communication sciences. More radically, as the writings of historian Jennifer S. Light and Marxists Éric Alliez and Maurizio Lazzarato suggest, it questions the possibility of distinguishing warfare and welfare in industrial democracies.[9] Progressive Era and interwar programs to reform society by means of science and technology, based on the assumption that scientific invention would lead to social reform, laid the foundations for a vast transformation of warfare itself. In making scientists, experts, policymakers, and foundations the most prestigious and well-funded instruments for reordering the polis, scientific philanthropies carried on an administrative struggle against the forces of social disruption—poverty, racial and ethnic antagonism, international warfare, the labor movement, anarchists—that threatened industrial capitalism and its oligarchic beneficiaries. In this respect, media theorist Daniel Nemenyi's claim that "the distinction between peace and war has never held ground with respect to the internet" is in fact true for a much wider apparatus of media sciences, including information theory, cybernetics, game theory, and digital computing, which took primitive form under the aegis of robber baron philanthropies.[10] So-called peacetime funding that gave impetus to this research, typically channeled via elite institutions such as Harvard, MIT, and Columbia, amounted to an administrative sally against the enemies of capital. These reforms, first felt in the fields of science, policy, and social welfare, gained prominence in the 1940s in military research, which projected their influence into the policies and science of the Cold War.

Robber Barons and the Technical Reordering of Science

Weaver's boosterism for informational analysis supported the Rockefeller Foundation's longstanding programs of global social reform. The foundation's tremendous success in reforming fields such as public health stemmed, in part, from its encouragement of techniques of experimentation, precision, repeatability, instrumentation, quantification, and economy among its beneficiaries. A peculiar interest in theoretical and formal models, often drawn from technical and bureaucratic systems, also featured prominently in its funding agendas. Its methods had roots in the Progressive Era, when many liberal Americans came to believe in science, expertise, and technology as neutral means for redressing the shortcomings of global capitalism. Against perceived threats of anarchy and communism, with their prospects of reassessing property rights and the distribution of wealth, progressives embraced the gradualist reforms of scientific expertise. Experts, typically hailing from the middle and upper classes (though not their highest echelons), embraced ideals of meritocracy and professionalism that moderated direct demands for radical political change by means of the bureaucratic and administrative apparatuses prevailing in modern research institutions.[11]

Scientific philanthropies, including the Rockefeller, Josiah Macy Jr., and Ford Foundations, led the way in championing a style of technical expertise that favored the eventual rise of communication research in the human sciences. Their midcentury interest in fields such as cybernetics and information theory reflected progressive hopes to submit divisive political issues for neutral technical analysis. Funded by Gilded Age magnates such as Andrew Carnegie and John D. Rockefeller, these organizations embraced a strategy of allying private capital with the scientistic administration of social affairs. The term *scientific philanthropy* referred to their organizational principle, according to which they dispensed funding on a rational basis rather than as faith-based charity. These institutions were typically dominated by professional experts who often had backgrounds in law or advanced degrees in the sciences, and they gave special priority to institutions of technical expertise. This approach promised to redress inequalities tied to race, class, and ethnicity through expert reforms that left existing property relations intact.[12] Carnegie's preferred institutions, such as libraries, hospitals, universities, and galleries, were staffed by experts and technicians; these institutions doled out measured quantities of education, edification, and social services while managing care for the public. The result was a flexible apparatus that took public and political problems from suffering to inequity and turned them into semiprivate and technical problems to be handled by an array of experts. This apparatus aimed to cultivate a class of entrepreneurial, rational, industrious citizens who yielded to the ministrations of experts and not to the call of radical politics. In turn, the political subjects produced by this apparatus were to become productive workers contributing to the maintenance and expansion of a rational, enlightened industrial society.

Historians credit Carnegie with founding scientific philanthropy, but the Rockefellers perfected it.[13] They established an unrivaled apparatus of administrators, institutions, endowments, public-private partnerships, and expert givers and receivers of funding, all of which remade philanthropy and policy as well as the personal and collective practices (educational, hygienic, agricultural, etc.) of populations throughout the United States and abroad. The family of Rockefeller philanthropies had expended more than $821,000,000 on research and public programs by the mid-twentieth century.[14] Its main institution, the Rockefeller Foundation, first launched in 1909, supported "the acquisition and dissemination of knowledge, in the prevention of and relief of suffering, and in the promotion of any and all of the elements of human progress" in the United States and around the world.[15] Its programs placed particular emphasis on practical problem-solving through scientific expertise and on the cultivation of that expertise around the world to promote an international climate of progressivism, tolerance, and openness.[16] Techniques for this transfer included reliance on external funding bids, an emphasis on teamwork, and a favorable disposition toward bureaucratic and administrative methods.

A number of scientific philanthropies turned their attention to new media technologies as well. Officers at the Rockefeller Foundation and the partner organization it ultimately subsumed, the Laura Spelman Rockefeller Memorial, described film and broadcasting as key "instrumentalities" for "formal education and for the general diffusion of culture."[17] In a similar spirit, they regarded social sciences as "instrumentalities for attainment and diffusion of knowledge" and history as "a technique of operation rather than a body of knowledge."[18] These perspectives converged in a general thesis of science as technique, whose refinement contributed to the fashioning of stable "social technology."[19] Fosdick, who served as president of the Rockefeller Foundation from 1936 to 1948, articulated the links between communication, technology, and social ills that would lead foundations and liberals to embrace cybernetics. Like many progressive intellectuals of the period, he saw communication and technology as central to the problems of interwar Western society. In a series of speeches under the title "The Old Savage in New Civilization," he cast the problem in terms of a clash between progressive modernity and benighted primitivism. "Our machine civilization," he declared, "has wired the world together in a vast, intricate circuit; the electric spark that starts anywhere on the line will travel to the end."[20] In this vision, civilization did not merely use machines; civilization became a machine—a "vast, intricate circuit"—subject to the laws and mechanics of networked machines. Research into communication provided dual enlightenment: into the technical sciences governing machinery and into the strategies for its political mastery.

A belief that science should pivot around technical research and rational communication guided officers at the leading robber baron philanthropies. This ideal had deep roots in Anglo-American experimental methods, according to which physical experimentation provided a neutral method for resolving contentious differences. In a 1933 memorandum on the benefits of science, Weaver proclaimed, "There is no more effective enemy of passion and prejudice than the calm temper of the scientific mind."[21]

Within a given society, it promoted salubrious and tolerant skepticism among its adherents. Globally, it secured the "international friendliness and understanding that results from a worldwide fraternity of scientists with their unifying bond of impersonal and unselfish interest and understanding."[22] Weaver's vision of science found its origins in Robert Boyle's seventeenth-century ideal of science as an honorable practice carried out by gentlemen who, equipped with the proper experimental and technical means, produced truths in peaceful retreat from the contentious public sphere.[23] In this account, instruments and experiments served as not only demonstrations of facts but also as impartial means of communication on the basis that networks of exchange reaching well beyond the lab could be established. However, where Boyle turned to judgment, witness, and eloquence as sources of an impersonal bond among scientists, the Rockefeller Foundation followed a trend in nineteenth- and twentieth-century liberalism that embraced the rationality and orderliness of modern technologies as the guarantors of reason.[24] Officers referred to legal systems, mass media, governments, social welfare, and hygiene services as types of "social technology" and likened social scientists to engineers whose task it was to identify and develop mechanisms for their control.[25] Rockefeller social science substituted technologies of control for political strife and class difference as well as for the scientist's powers of judgment and witnessing.[26]

The robber baron philanthropies' deference to technical research generally, and to industrial communication specifically, also reflected a change in the sites of knowledge production in the twentieth century. Communication theorist James Carey noted communication industries' key role in scientific and political thought in the United States.[27] As the first great industrial monopoly in the United States, the Western Union telegraph company provided a model for politicians and businessmen alike of a certain kind of industrially managed universality. The success of that monopoly furnished a new model for how to conduct science in liberal capitalist democracies. In the wake of telegraphy, the values of large-scale industrial organization increasingly inflected scientific research. Professional engineering societies, institutes of technology, and research laboratories allied technological research with a rhetoric of civic responsibility. These organizations, staffed by educated white-collar professionals, furnished actionable research for industrial society. Philanthropies' support for research on technical communication amplified these organizations' identification of an orderly modern state with a well-engineered modern state.

A contrast with European sciences indicates the peculiarity of American robber baron philanthropies' emphasis on technical communication. As fascism drove European intellectuals to seek political refuge in the United States, US philanthropies quickly enlisted them in experimental programs of technical research. Consider the role of Weaver and the Rockefeller Foundation in shaping the agendas of the Unity of Science movement as its center of activities shifted from Europe to North America. The movement embodied one instance of a broader interest in communication among interwar intellectuals. Eminent members such as Otto Neurath and Rudolf Carnap of the logical positivists in Vienna sought to establish a scientific language that would allow for unambiguous communication across cultures.[28] Believing that misunderstanding was a principal source of global conflict, they turned to methods in logic, physics, and even iconography to forge cross-cultural and interdisciplinary understanding. As members resettled in the United States, however, technology increasingly took the place of logic and reason. For example, in 1935 an ally of the movement, American philosopher Charles Morris, appealed to Weaver for support of efforts to establish a philosophical language that bridged disciplinary and linguistic contexts. In response Weaver commended the group and made two suggestions. First, he suggested that "many of the subtle and perplexing problems of language and of meaning are perhaps best approached through the use of . . . logical machinery," a reference to the plans for a Rockefeller-funded differential analyzer at MIT, which was among the first modern electronic computers.[29]

Second, Weaver advised Morris to contact C. K. Ogden and I. A. Richards of the Basic English project, which sought to establish English as the standard language for communication worldwide. An acronym of British, American, Scientific, International, and Commercial, the inventors of Basic English advertised global standardization indexed to technical and scientific methods. The project benefited from Rockefeller Foundation funding, particularly in promoting its introduction to curricula in China.[30] It involved an effort to economize expression, much as information theory and other postwar methods would. Aided by statistical analysis of English, the inventors of Basic had identified commonly used English words and devised a combinatory matrix for assembling complex meanings out of elemental combinations. This method attracted widespread interest among mathematicians and engineers, including Weaver, who had written on the mathematics of English vocabulary in the 1920s, and Shannon, who cited Basic English in his efforts to define the probabilities governing English vocabulary.[31] Inexpensive pamphlets presenting works by the likes of James Joyce in Basic English allied this technoscientific method with economies permitted by industrial printing.[32] Basic English modeled an instrumental approach to meaning-making, complementary yet distinct from differential analyzers, at home in a larger family of techniques for the statistical and logical profiling of language. With Rockefeller support, all symbolization seemed rife for improved technical processing.

Weaver's recommendations must have struck Morris, the cosmopolitan logical positivist, as bizarre; it was not at all clear how the mechanical wheels and rods of 1930s computers might be adapted to establish broad scientific principles cutting across the natural and human sciences, much less lead to the kind of intercultural and political cooperation envisioned by the Unity of Science movement. In the years after World War II, however, leading members of the Unity of Science movement resettled in the United States, entered into loose partnerships with Wiener and his collaborators, and, by the late 1940s, embraced cybernetics and informatics as key sciences in the effort to develop a universal language of scientific communication. Upon doing so, they won funding from Weaver and the Rockefeller Foundation to rebuild their movement in the United States. Within this process of epistemic reform, Rockefeller funding supported new concepts and lines of investigation, for example, by nudging the Unity of Science adherents in the United States to embrace cybernetic and informatic paradigms. Historian of science Peter Galison argues that the embrace of cybernetics by the Unity of Science movement reflected the effects of a World War II mobilization that transformed the nature of science, putting cybernetics, radar, and other technical sciences at the center of the scientific agenda.[33] The exchange with Weaver, however, indicates the priority of interwar and even early twentieth-century agendas of robber baron philanthropies in these intellectual accommodations.

The Interwar Affinities of Computing at MIT

The Unity of Science movement in America illustrates a trend in the postwar ascent of communication science. While World War II offered the most proximate scientific and technical event for the rise of enthusiasm for the communication sciences, dense multilateral networks among technologies, institutions, foundations, and intellectuals formed its basis. Links among the Rockefeller, Macy, and Carnegie philanthropies forged in the

1930s and 1940s, well before the United States' entry into World War II, guided subsequent initiatives in cybernetics, information theory, and game theory. Robber baron initiatives interwove social and scientific reforms with technical methods that seemed to promise new precision and insights into for cultural analysis. This was particularly the case in the funding programs for computation and informatics, which from their earliest development linked up with a family of problems in industry, aesthetics, biology, criminology, and other fields. Over the course of decades, marked at key moments by coordination with the national military and intelligence aims of the US government, champions of computing such as Vannevar Bush of MIT and Weaver cultivated a flexible apparatus of findings, researchers, institutions, applications, and instruments that positioned the field for its postwar blossoming into what Geoffrey Bowker termed a "universal science" and what Ron Kline called an "umbrella discipline."[34]

Consider the milieu of the aforementioned Rockefeller differential analyzer, envisioned by MIT professor Vannevar Bush as the centerpiece of an MIT-based Center of Analysis that would put mechanical analysis in service of diverse scientific, industrial, bureaucratic, and governmental purposes.[35] Bush's vision for the device encompassed social, political, and scientific problems bridging mathematics, criminology, eugenics, and industrial consulting. As historian Colin Burke has shown, Bush conceived of the center as a site for integrated research and development of machinery in electronics, photoelectricity, automated memory devices, and computing as well as in "digital calculation and machines to solve the escalating problem of file management in science and bureaucracy."[36] In the development of the Rockefeller Analyzer and the center, Bush mapped out a field of relations that cut across physics, sociology, mathematics, and industrial problem-solving. It also anchored a set of social relations, as development of the device served as a platform for consultation and collaboration with Wiener, Weaver, the Rockefeller Foundation, and the Carnegie Institute. Appealing to Weaver for funding, Bush championed the extraordinary promise of electronic computing as an interdisciplinary device "to which research workers everywhere will turn for their solutions of their equations."[37] A distinct alliance of universities, philanthropies, and industry also defined the analyzer and its proposed center. In justifying the electronic design of the Rockefeller Analyzer, Bush argued that time saved by foregoing labor-intensive processes of configuring analog components would free up time on the machine that could be rented to industry to defray costs.

This network of affinities formed around the analyzer—social, personal, economic, political, scientific, industrial, educational, institutional—convinced Weaver and the Rockefeller Foundation to sponsor its construction. Bush's expansive vision launched an early conception of informatics as a universal science, suitable for the standard processing of word, image, and sound, adapted to the analysis of natural as well as cultural facts. Precisely because of the Rockefeller Foundation's aversion to funding research into narrowly applied technologies, this vision of broader scientific and social instrumentation met the conditions required to open its vast coffers. To grasp the implications of Bush's proposal, consider his computing and simulation devices before the Rockefeller Analyzer. Analog devices such as the Network Analyzer and Product Integraph simulated real, embodied physical relations. In Weaver's words, their analog mechanisms conveyed an instructive "vividness and directness of meaning," in contrast to which "a Digital Electronic computer is bound to be a somewhat abstract affair, in which the actual computational processes are fairly deeply submerged."[38] That abstraction, however, opened up infinite modularity, by which the real distinctions among objects became relativized in strictly symbolic forms. The Rockefeller Analyzer initiated that new discursive order of pure simulation untethered from the real. The first user to demonstrate full appreciation was Shannon. He had been hired as a graduate student to act as operator for the analyzer and published his findings in his 1938 landmark master's thesis, "A Symbolic Analysis of Relay and Switching Circuits."[39] Shannon demonstrated that the setup of computing circuits could be designed and tested according to algebraic procedures presented by philosopher George Boole in his nineteenth-century treatise *An Investigation of the Laws of Thought*.[40] Shannon's analysis offered the first proof of the logical programmability of computers according to principles guiding programming today.

The interest in Shannon's analysis stemmed, in part, from its success reducing a process of messy, trial and error problem-solving into an orderly logical series. With the aid of Boolean algebra, the many choices and operations executed by the Rockefeller Analyzer could now be defined in terms of binary selections between discrete, well-defined terms, reflecting the binary design of the relay circuits in the Rockefeller Analyzer. Shannon's presentation offered a landmark paradigm for rigorous analysis: intuition and observation gave way to the technical precision of unambiguous instrumentation. Problem-solving became, literally, a problem of "relays," according to a series of well-defined logical operations. This hardly dispelled the great mass of intuition and observation necessary to program the solution of a mathematical problem. Even so, Shannon's thesis put forth a bold proposition for the automation of messy scientific problems. He saw that much of the work animating everyday scientific work might be subject to mechanical solutions. His subsequent essay, "Mathematical Theory of the Differential Analyzer," suggested an identification of the machine with theory itself, wherein electricity, switches, and rods corresponded to idealized theorems.[41] Set in the context of Bush's scientific, industrial, bureaucratic, and governmental ambitions for MIT's Center of Analysis, Shannon's work implied the question of which problems—if any—lay outside the scope of technical problem solving. Perhaps the very laws of thought that Boole had in mind when he developed the algebraic notation used by Shannon could be implemented in mechanical computers.

Bush's and Shannon's ambitions for symbolic automation permeated their work on a second information processing device, the Rapid Selector (figures 1.1 and 1.2). In that project, political prospects for symbolic analysis came to the fore. In a manner characteristic of Progressive Era technocracy, in which the scientific processing of records seemed to be the linchpin of an orderly polis, Bush and his colleagues envisioned the rapid selector as a general-purpose information processor suited to the analysis of genetic, business, library, and statistical data.[42] Throughout the 1930s and 1940s, the Rockefeller Foundation had provided lavish support for the development of microphotography, a forerunner of microfiche, for documentation in the humanities. As such, the foundation stood out as an excellent candidate to support Bush's work with the selector. Weaver recorded in his notes from a 1937 meeting, "Sometime ago federal authorities asked him [Bush] to consider the problem of devising mechanical aids for rapidly locating finger prints. . . . B[ush] worked out a system which would permit the examination of approximately 1,000 per second." Considering the problem of rapidly sorting fingerprints, Weaver wrote, "it then occurred to him [Bush] that this scheme was possible for development into a new technique for making available the stored literature of the past."[43] In this manner biological and bibliographical data might be processed by the same techniques of automation exploited by mechanical computers.

Bush's initial accounts of the Rapid Selector emphasized the storage of scientific publications, but subsequent accounts expanded beyond biopolitical criminology to include legal, historical, and ethnographic data. In a much-celebrated elaboration that appeared in 1945 in the popular magazines *Atlantic Monthly* and *Life*, Bush presented his devices as a general research instrument he termed the Memex: essentially a desk equipped

1.1 and 1.2   Illustrations of the coding and processing mechanisms of the Rapid
Selector, taken from an improved version of the device based on
Bush's design. The machine-coded microfilm permitted automatic
sorting according to binary on-off sorting mechanisms triggered
by the opacity or transparency of gridded segments. Source:
John C. Green, "The Rapid Selector—An Automatic Library,"
Military Engineer 41, no. 283 (1949): 351.

with microfilm sorted by numerical codes (figures 1.3 and 1.4).[44] Updating Boole's vision for the laws of thought, Bush explained that with the Memex, "simple repetitive thought could be done by machine, following the laws of logic."[45] Research and analysis would be recorded by machines and turned into algorithms that future research could reenact. He provided the example of a would-be researcher reconstructing the archaeological history of the bow and arrow. "Specifically," Bush explained, "he is studying why the Turkish bow was apparently superior to the English longbow in the skirmishes of the crusades."[46] After a deep dive into the academic literature, he takes notes, and his inquiries and commentaries are encoded by the Memex. "Several years later, his talk with a friend turns to the queer ways in which a people resist innovations. . . . [In response,] a touch brings up the code book" on the Memex, and he can quickly retrieve the entirety of his previous inquiries as an algorithmically defined discursive series.[47] It was much more than a card catalog and quite different from a technical predecessor to the World Wide Web. Bush was elaborating a systemic theory of branching (or dendritic) analysis, applicable to words, numbers, images, and even biological markers, according to which technology not only calculated but also indexed order, hierarchy, and interrelation among "contents." Perhaps natural history could, through sufficient indexing, succumb to a logical and formal description, according to relations more profound than those produced by human consciousness.

Ethnography and archaeology furnished Bush and his colleagues with patterns of culture for informatic reconstruction on other occasions as well. In 1953, Bush and Shannon spoke alongside anthropologist Earnest A. Hooton, who was well known for his research in racial classification.[48] Together they offered prognoses of the position of atomic power and computing in the long arc of human progress. In another public address, Shannon explained that "modern information processing and communication" continued a "great stream of technological activity" inaugurated when "early man learned to communicate with his fellow man by the spoken word."[49] This line of analysis was common in the period, with scientific introductions to information theory often linking it to problems that arose with primitive man and the growth of spoken and written languages.[50] One early information theorist, Colin Cherry, identified the rise of communication theory with the rise of phonetic European writing. He cited early Mediterranean and Egyptian scripts as highly redundant logographic texts, rife with inefficiencies, that gained economy and manipulability via phonetic writing and Roman shorthand.[51] Through these kinds of analytical

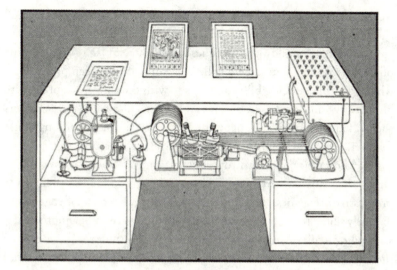

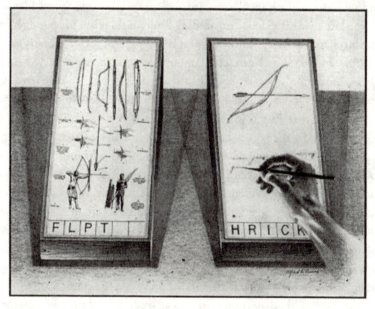

1.3 and 1.4   *Life* magazine illustrations of Bush's account of the Memex, with the Rapid Selector embedded in an office desk. Where the first image appears to show geological research (note the figure of a volcano on the display), the second offers a visual illustration of the history of the longbow paired with handwritten commentary by the user, coded by text at the bottom of the interactive screen. Source: Vannevar Bush, "As We May Think," *Life*, September 1945, 123, 124.

frames, thinkers such as Bush, Weaver, Wiener, and Shannon indicated the fundamental anthropological significance of communication sciences. As French philosopher Jacques Derrida would later note, this line of analysis also recapitulated a hallmark trope of European philosophy, which identified intelligibility and reason with the emergence of discrete phonetic writing systems in Europe, and which culminated in researchers such as François Jacob arguing in the 1960s that the basic building of life followed principles of informatic and cybernetic inscription.[52]

The biopolitical and anthropological affinities around computing at MIT in the 1930s reached into one of the great passion projects of interwar progressives: eugenics. Bush apparently viewed the reform of eugenics as uniquely suited to the informatic formalism innovated by Shannon's analysis of relay circuits. Around 1939 Bush left MIT to assume the presidency of the Carnegie Institution, another of the great robber baron philanthropies and a sponsor of the Eugenics Record Office (ERO) in Cold Spring Harbor. By the late 1930s the deep and abiding biological racism propounded by the ERO had fallen out of fashion, and Bush sought to reset it on rigorous scientific foundations. Bush seized upon Shannon's theory of relay switching as a model for how a haphazard, unscientific method could be rendered orderly and logical with applied mathematics. Bush later explained, "It occurred to me that, just as [Shannon's] special algebra had worked well in his hands on the theory of [binary] relays, another special algebra might conceivably handle some of the aspects of Mendelian heredity."[53] Writing to ERO psychologist Barbara Stoddard Burks, who had mined the ERO's sprawling records for evidence of a genetic basis for intelligence, Bush explained that he had suggested to Shannon that he "try his queer algebra" in the analysis of genetic problems.[54] This offered a chance for Bush, who had intermittently championed bioengineering over the past decade at MIT, to realize his efforts to bring logical and computational analysis to bear on an infamously complex problem deemed of pressing social importance.

Shannon's PhD dissertation in genetics, completed in 1940, sought a mathematical solution to a deeply political problem: the orderly processing of eugenic data. Bush's long-standing goals for streamlined information processing methods figured in his belief that Shannon should work with the ERO. There, biological racism and shoddy records appeared as obverse sides of a single devalued coin. The title of the infamous research center, Eugenics Record Office, reveals the bureaucratic banality of its malign project. From the compilation of vast records on kinship, particularly in families supposedly marked by alcoholism, mental retardation, dementia

praecox (schizophrenia), and "wanderlust," its administrators would develop techniques for purifying America's human stock (figures 1.5 and 1.6).
The ostensible chasm between eugenic and modern-day ethnographic
approaches to kinship (with their respective emphases on linear heter-
onormative transmission and multiple attachments) belies their history of
intersecting techniques. In the nineteenth century, eugenicist Francis Galton strove to reform kinship diagrams with the aid of the "binary system
of arithmetic" founded on the simple distinction between zeroes and ones
(figure 1.7).[55] Despairing that even scientists were unfamiliar with binary
notation, he admitted that other systems would be necessary for the time
being. Even so, his efforts to introduce algebraic rigor to kinship diagrams
attracted interest from some ethnographers. By the 1930s, however, eugen-
icists' efforts to turn their data into formal kinship diagrams was widely
recognized as deficient. A Carnegie audit of the ERO called the immense
system of eugenic records at the ERO a "vast and inert accumulation" that
was "unsatisfactory for the scientific study of human genetics" and characterized its "system of recording" as "unsound."[56] Bush used these findings
to force eugenicist Harry Laughlin, a leading administrator at the ERO,
into early retirement.[57] Parallel to forcing a biological racist out, Bush,
who controlled the purse strings that would decide the ERO's fate, urged
one of its researchers to accept Shannon as a student. Viewing the kinship
notations adapted by the ERO from ethnography, Bush may have seen in
the unwieldly diagrams a system of binary relays not unlike that of the
differential analyzer: a multitude of complex surface phenomena pivoting
around the integration and derivation of parental pairs. Shannon's "queer
algebra" boded the possibility of an elegant mechanism for classifying records and explaining genetic transmission, much like he would devise for
telephone relays and chess-playing computers (figure 1.8).

   More crucially, the analytic endeavors growing from the Center of
Analysis indicate how foundations, universities, the state, and industry
collaborated in technocratic reform. The ascent of technically minded
expertise, instrumentation, scientific notation, institutions, networks of
insiders, and computation occurred through decentralized, ad hoc coordination. Foundations such as those established by the Rockefellers and
Carnegies acted as relays for financing, expertise, and personnel, with the
power to realize or halt lines of investigations and instigate new ones (as
when Bush forced Laughlin out of the ERO and brought in Shannon). The
piecemeal nature of these reforms precluded sweeping epistemic decrees.
Rockefeller Foundation support for the differential analyzer produced

1.5 Eugenics Record Office archives room with card index on far wall and field worker files on right, ca. 1921. Source: Image 1649, Cold Spring Harbor Laboratory Archives, Truman College.

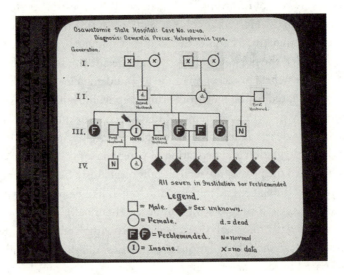

1.6 "Osawatomie State Hospital: Case No. 10240. Diagnosis: Dementia Precox, Hebephrenic type." A "pedigree" record from the Eugenics Record Office concerning the history of "normal," "female," "feebleminded," and "insane" family members of a woman institutionalized in a Kansas hospital with a diagnosis of schizophrenia (or "dementia praecox" as it was more commonly known), ca. 1925. Source: Image 920, Harry H. Laughlin Papers, Truman State University.

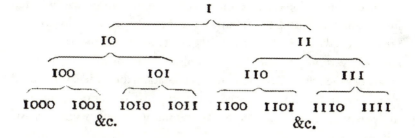

1.7   Eugenicist Francis Galton's 1883 proposal for a binary arithmetic
for kinship. He used the opposition of zeroes and ones to notate
the networked relay of qualities across generations, including
an encoding of the multiple attachments that obtained across
branches. Source: Francis Galton, "Arithmetic Notation of Kinship," *Nature* 28, no. 723 (September 1883): 435.

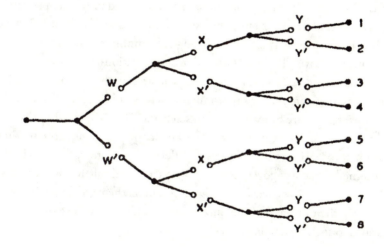

1.8   Claude Elwood Shannon, "Disjunctive
Tree with Three Bays" (1949). Source:
Clause Elwood Shannon, "The
Synthesis of Two-Terminal Switching
Circuits," in *Claude Elwood Shannon:
Collected Papers*, ed. N. J. A. Sloane
and Aaron D. Wyner (Piscataway, NJ:
IEEE Press, 1993), 597.

landmark innovations and new intellectual communities that persisted in MIT and computing for decades. By contrast, Shannon's dissertation proved a dead end. After a successful defense he abandoned the work to return to wartime problems in communication engineering. Yet, over the course of generations of funding and collaboration, these endeavors seeded elements of a common analytical framework capable of bridging inquiries in fields as far-ranging as electronic computing, eugenics, and forensic criminology.

In time, the philanthropies' policies cultivated a network of affinities that allied diverse fields with informatics. The effort to win funding from Rockefeller or Carnegie philanthropies drove researchers of the period to envision laboratories, centers, and applications that went well beyond narrow academic problem-solving. Initiatives that promised to submit difficult problems to machine-based processing excited particular interest, appealing to the philanthropies' bias toward technical problem-solving. On this basis, computing and information processing emerged as objects of concerted interwar interest, gathering together the likes of Bush, Wiener, Weaver, and Shannon in an integrated program of research and education. Foundation sponsorship also encouraged hybridization, as when Bush's Carnegie affiliation brought Shannon into the ERO and its eugenic initiatives. The elite positioning of foundations, moreover, allowed their privileged researchers—engineer-scientists with an entrepreneurial bent—to instigate reforms reproducible across public and private institutions, from state bureaucracies to scientific laboratories. They sought out techniques for reducing complex epistemic problems, from sorting criminal forensics to the solution of differential equations, into logically patterned sequences governed by definite series and calculable selections. While these projects met with early success in mathematics and industrial research, their promoters remained optimistic about their eventual applications to social research.

Cybernetics as the Study of Man

The hybridization of the human sciences with technocratic engineering expertise figured even more prominently in the genesis of the celebrated Macy Conferences on Cybernetics (initially convened in 1942 as the Cerebral Inhibition Meeting, reorganized in 1946 as the Feedback Mechanisms and Circular Causal Systems in Biology and the Social Sciences

Meeting).[58] The patron of cybernetics, the Josiah Macy Jr. Foundation, promoted social reform through medical research. In keeping with Progressive Era theories of social hygiene, its officers viewed biology and welfare as part of public health. The cybernetics conferences belonged to that program of social hygiene. Macy officers saw the conferences as a forum where the rational tasks of the humanities and social sciences could be developed and gradually extended to the natural sciences. This notion of extending techniques from the human sciences to other disciplines nearly inverts latter-day interpretations of the conferences as showcases for informatics and game theory. Frank Fremont-Smith, Macy Foundation medical director, explained at the opening of the 1949 meetings, "The problem of communication is largely a problem of human relations and for its solution requires intensive and comprehensive scientific study of man. In order to study man it is necessary to bring in every one of the physical and biological sciences and every one of the social sciences also. . . . *Thus in the study of man we may find eventual unification of all the sciences.*"[59] Cybernetics was a humanism, demanding collaborative study from the likes of Rosenblueth, Wiener, and Shannon as well as from anthropologists Margaret Mead and Bateson. Between 1940 and 1950, in its effort to promote human health and well-being, the Macy Foundation sponsored 132 interdisciplinary conferences with some 800 participants.[60] These conferences spanned anthropology, linguistics, biology, engineering, and physics in the hope that communication between disciplines would generate scientific unity proper to the "study of man." The technicity of cybernetic humanism, its emphasis on synthesis through technical systems over individual or historical experience, sprang from the same technocratic spirit that infused the robber baron philanthropies.

In 1949, in homage to Wiener's book on cybernetics, arguably the first major publication resulting from the conferences, attendees unanimously agreed to add *cybernetics* to the name of the feedback conferences, and the meetings became popularly known as the Macy Conferences on Cybernetics.[61] This was a nice gesture but also quite practical, given the inscrutable mouthful their original title presented. Subsequent accounts then erroneously took that gesture as crediting Wiener with a project that had been initiated by other researchers and underway for years. Preliminary conference discussions took place in 1942, without Wiener. The group proceeded from 1946 onward with his participation, but the overarching emphasis was on interdisciplinary dialogue across social and biological sciences. In 1948 this resulted in Wiener's book *Cybernetics*, in which he emphasized its roots in collective discussions that included Mead and Bateson and were sustained by institutional patrons including the Macy, Rockefeller, and Guggenheim Foundations.[62] The overarching framework was conceived to allow for mathematical contributions such as those offered by Wiener. The originality of Wiener's synthesis under the name *cybernetics* concerned its repackaging of these themes under a loose framework of communication engineering. Yet the themes of that treatise, spanning engineering, philosophy, psychopathology, and ethics, were notably at odds with his previous writings. As an epistemic ensemble it reflected the agenda of Macy Foundation officers and their social-scientific confederates, not Wiener (and it most certainly did not reflect the actual contents of his short-lived, failed artillery-control research during World War II).

Yet Wiener's synthesis offered a great service to the technocratically minded framework in which the likes of Fremont-Smith and Mead sought to establish their significance. The imprimatur of a celebrated mathematician with bona fides in electronic computing like Wiener, hailing from MIT, allowed Mead and her colleagues to ingeniously repackage their interests in the ostensibly exact terms of natural science and engineering. Thus a problematic cultivated in the human sciences and given shape by foundation administrators won legitimacy in an intellectual context that celebrated engineering knowledge. In particular, Mead and Bateson recast anthropologists' claims that societies might be described in terms of what ethnographer Ruth Benedict terms "patterns of culture" in the jargon of data transmission borrowed from cybernetics and information theory. A bit like the cuckoo that lays an egg in the nest of another species for rearing, Wiener, never known to suffer from false modesty, seems not to have recognized the extent to which his research program became the nest for bringing into the world concepts cultivated by his peers in decades prior. Said concepts' sponsors had hoped that a properly technical and scientific framing of social problems might clarify the pathologies of third-world colonies as well as those of the urban slums of Western industrial nations. Subtly, critically, and imaginatively, psychologists, psychoanalysts, and linguists turned to mathematics and engineering to recast their own findings in a framework that would be sufficiently technical to win wider acclaim in postwar American science.

Lawrence Frank's role convening and directing a network of conferences related to cybernetics clarifies the agenda behind their study of man. From the 1940s through the 1960s, Macy conferences covered topics from hypnotism to LSD as part of a constellation of ideas linked to feedback and communication. The main Macy-funded cybernetics conferences ran from 1946 to 1953, but key members of the group began meeting as early as 1941 or 1942 under the energetic leadership of Frank, a vice president at the foundation. He brought together key members, including neurophysiologist Warren McCulloch, Rosenblueth, Mead, Bateson, and psychiatrist Lawrence Kubie, in the study of cerebral and feedback mechanisms.[63] (Frank set in motion and participated in the conferences that were overseen by his mentee, medical director Frank Fremont-Smith). After undergraduate studies in economics at Columbia University, Frank worked for the New York Telephone Company and later the New School for Social Research. In 1923 he accepted a position as an administrator for Rockefeller philanthropic programs. He continued that work, in various roles, until 1936, when he accepted an appointment at the Macy Foundation.[64] His greatest gift lay in subtly crafting neither theories nor texts but communities and even generations of thinkers. Anthropologist Mary Catherine Bateson (daughter of Mead and Gregory Bateson) said that while Frank is best known "as the founding father of the child-development movement," his greatest work was as one of the "outstanding intellectual entrepreneurs of this century," responsible for creating new intellectual networks and conceptual seeds given fuller elaboration by other scholars.[65] Accounting for Frank's tremendous influence, historian of science Heims noted that as an executive of a well-funded foundation "answerable only to his board of trustees," he successfully induced "universities to shift the direction of their research along lines he advocated."[66] This gave men such as Frank and Weaver (nearly all the foundation executives were men) even greater liberties to shape research than administrators of the parallel programs of war-related research. Freed from the task of defeating the Axis, they could narrowly target research in areas poised to ripple across the university and domestic policy.

Frank gave early shape to cybernetics' preoccupation with kinship as a regulatory mechanism. His interest in kinship, a decidedly anthropological problem, partly sprang from his friendship with Mead. In the 1940s Frank and Mead merged households in a project of collaborative child-rearing. Their shared interests in family preceded that unusual arrangement, however. In the 1930s Frank began adapting elements of the Culture and Personality school of anthropology, with its emphasis on cultural patterning as a source of personality, to the study of the nuclear family (for more on this theme, see chapter 2). In his 1940 article "The Family as a Cultural Agent," a hallmark of the cybernetics meetings appears in rudimentary form: the belief that society and culture are organized as highly patterned systems that act as control mechanisms on individuals. Echoing Mead, he argued that families transmit these patterns to children, who in turn reproduce the patterns of the entire community. The goal of identifying these systemic properties was improved mental and social hygiene. Frank explained that "unhappy, distorted, anti-social, individuals, engaged in activities that we call delinquency, crime, sex offenses" reflected a breakdown in these cultural patterning techniques.[67] This view of culture, understood in terms of communication systems and control mechanisms and oriented toward the transmission and relay of consistent patterns among its members, spread through the Macy conferences. It envisioned children, parents, play, mealtimes, and routines as heterogeneous elements of a common of infrastructural relay. With the successful transmission of patterns, it suggested, society could count on forming an upstanding and productive citizenry.

Frank's 1940 account of the family as a cybernetic apparatus that "transmits the culture to the child, [and] attempts to socialize him," set the epistemic stage for the abiding interest in "pattern" at the Macy conferences. Under his steady administrative hand, an ethnographic meaning of pattern that was rooted in kinship theory merged with communication engineers' notions of pattern as the meaningful index of a signal economy. To be sure, accounts by McCulloch of neurons emitting "complicated sequential patterns," by Wiener of the "statistical pattern of the data" in his antiaircraft devices, and by Kubie of neurotic illness as "pattern of distress" had discipline-specific meanings.[68] At the conferences, however, *pattern* took on a decisive meaning for the study of man, laying the groundwork for the aforementioned claims that information theory might account for social patterns within primitive and prehistoric life. Frank thereby helped define the intellectual framework that allowed Wiener's cybernetics and mathematician John von Neumann's theories of games and computing to align with human scientists' theories of cultural order. Gradually, a new cultural analytic took shape around this palimpsest of ideas. An ethnographic theory of cultural patterning, developed in the study of colonial subjects and ostensibly dysfunctional American families, provided the model for the integration of information theories and social theories.

If Frank framed the cultural and social significance that engineering research would have in postwar science, it is nonetheless the case that the arrival of mathematicians and engineers decisively shifted the discourse of social scientists in the Macy group. Social scientists' deference to participants with authority in engineering created hierarchies that legitimated the ideas of the former by reference to the claims of the latter. This conferred on technological research an enhanced political authority. Before the Macy conferences, research in computing and communication engineering had little to say about society writ large. Under the leadership of Frank and the Macy Foundation, with the coordinated efforts of Weaver and Bush and their foundations, communication engineering won for itself expanded authority for cultural analysis, which would eventually take the form of fully fledged informatic theories of social order.

The case of psychologist Molly Harrower, an expert in Rorschach tests and a founding member and regular attendee of the conferences from 1946 to 1949, speaks to the historical distortions that resulted from prioritizing engineering and mathematics in the history of the cybernetics movement. In her work we find a more compelling account of the concern with patterns that gradually spread through the cybernetics group. Harrower was a long-standing beneficiary of Rockefeller and Macy financing, as well a wartime consultant to the US surgeon general and the State Department.[69] An acquaintance of Frank since 1936 and of McCulloch for years prior to the feedback conferences, she exemplifies the long-term layering of agendas, usually oriented around scientific management of social pathology, that permeated scientific philanthropy. Her interest in poetry as a therapeutic method suggests new genealogies for understanding the later preoccupation of participants such as Wiener and Shannon with discursive patterning and pathology. The almost total disappearance of Harrower from histories of cybernetics speaks to the success of the conferences in founding the myth of mathematics and engineering as a rigorous engine for ideas disseminated to the softer human sciences. That myth reinforces the idea of a masculinist and wartime ontology underpinning cybernetics and set off from the humane interpretation of these ideas in the social sciences in years to follow. Well into the present day, the emphasis on the likes of Wiener and von Neumann in cybernetics occludes a history of early informatics in which female social scientists researching topics such as poetics and pictures figured prominently.

Emergence of a Cybernetic Apparatus

The affinities formed around the Center of Analysis and the Macy conferences in the 1930s and 1940s dominated the "cybernetics moment" of the 1950s and 1960s.[70] Roots sunk into eugenics, criminology, ethnography, and family psychology prefabricated the interpretive frameworks of postwar informatics. In that network a program of social control by technical means took shape, vouchsafed by a network of experts whose costs were defrayed by the wealth of robber barons. World War II provided a forum for the intensification of those ties, supported by even larger injections of cash and greater consultation with experts in fields such as physiology, biology, operations research, communication engineering, and even anthropology.[71] Faced with Blitzkrieg, kamikaze attacks, and indecipherable signals, the allies turned to information processing machines to translate words, sights, and nerve impulses into tractable data for mounting a counteroffensive. The US military turned to the philanthropies' networks of researchers to realize wartime innovation, forming intellectual compacts whose orientations owed as much to the Progressive Era and interwar technocracy as to the militancy of World War II. In the decades to follow, as cybernetics stormed the human sciences, the personnel elevated by prewar research would often continue as spokespeople for postwar research.

In June 1940 Bush, as president of the Carnegie Institution, presented President Franklin Roosevelt with a plan for a new organization "to conduct research for the creation and improvement of instrumentalities, methods, and materials of warfare."[72] His proposal launched the National Defense Research Council (NDRC), the arm of scientific research for the wartime effort. To oversee the mathematical division Bush appointed Weaver, under whom Shannon, Wiener, and mathematicians at the Center of Analysis conducted the research that outlined the contours of cybernetics, information theory, and digital computing.[73] Well into the 1950s, when foundations green-lit funding for interdisciplinary communication projects, they tapped into the networks cultivated in the 1930s and consolidated in wartime military projects. People who circulated across these programs—Weaver, Bush, and Shannon, together with the likes of John von Neumann, Wiener, Lazarsfeld, and Bateson—became the key figures in the reinterpretation of communication engineering for the human sciences. In this context, the interdisciplinary communication research of the wartime years, so often attributed to wartime exigency, is recognizable as the outgrowth of decades of science policy sustained by the technocratic alliance of foundations, industry, and universities formulated in the Progressive Era.

Long before the United States entered World War II in December 1941, Mead and Bateson pondered how their profession might serve the war effort—and, at least as decisively, facilitate a just postwar reconstruction. Because the first Macy Conference on teleology took place in 1942, historians often regard wartime mobilization as the prime mover in cybernetics—an interpretation promoted by Mead and Bateson, seemingly to win greater authority through prestigious association with engineering and science.[74] But their concern for the "systemically distorted communication" of modern life had clear roots in the work of Frank, Mead, and Bateson dating back to the 1930s.[75] As a Rockefeller officer, Frank sponsored Mead's work as far back as 1934, aiding her rapid ascent in New York intellectual life. Mead's marriage to Bateson in 1936 put the ideas of all three into contact, an exchange formalized when they began summering together in 1940. Their common interest in group dynamics emerged from social scientists' depression-era preoccupation with recasting social inequity in terms of maladjustment as a systemic problem. While all three rejected dogmatic US values around child-rearing and sexuality, a Progressive Era commitment to fostering social unity, national integrity, and cultural order generally infused their theories.[76] These orientations easily met two wartime demands: First, Mead and Bateson adapted their studies of indigenous societies to US-funded propaganda mobilizing US "morale" through canny magnification of internal group bonds. Second, Bateson applied these insights to propaganda disrupting or exploiting the cultural systems of enemy and occupied peoples, most notably in his twenty-month stint as a US military advisor in Ceylon, India, Burma, and China. To be sure, the eventual recruitment of Wiener, von Neumann, and Shannon to the Macy conferences benefitted from their shared experience as consultants in wartime communications research.[77] But, fundamentally, that alliance did not formalize wartime research so much as it crystallized ideas, on a larger scale, sketched by Frank, Mead, and Bateson since the 1930s. The success of the Macy conferences did, however, showcase how interwar social scientific expertise not only accommodated, but also captured, the burgeoning programs of World War II and Cold War communication research underway in the natural sciences and engineering.

Whatever social agendas shaped the prestige given to communication science after World War II, what they implemented was less an "idea" or a "theory" than a cultural analytic whose effects depended on relays forged among institutions, instruments, patrons, scientific inquiry, and social policies. The originality of the Macy conferences stemmed from their success propounding an approach to cultural theory rooted in interdisciplinary teamwork and theoretical analysis indexed to technology. The idea of the human sciences as interdisciplinary, theoretical, collaborative, and aided by empirical data figured prominently in decades of efforts by scientific philanthropies to overturn "armchair" philosophizing and speculation. Communication research and technical instrumentation had long been centerpieces of these efforts, as in the Rockefeller interwar communication research leveraging the teamwork of Lazarfeld, political scientist Harold Lasswell, anthropologist Geoffrey Gorer, and literary critic I. A. Richards, among others, to found a neutral scientific means of directing state propaganda.[78] Information theory and cybernetics offered these kinds of endeavors a path forward after the war, with a new emphasis on economy, fidelity, and error correction as measures of communication systems. These theories construed breakdown in terms of system failure rather than social contradictions. Mathematicians' and engineers' theoretical accounts of machines lent phantom neutrality to essentially political claims about the structure of communication. Thus a cultural analytic, defined by specific modes of collaborating, speaking, analyzing, prioritizing, and funding research, took shape, with communication engineering acting as an unimpeachable mouthpiece addressing postwar American social priorities.

Rockefeller's Epistemic Cleanup Crew

In the late 1940s and early 1950s, the Rockefeller Foundation (and Weaver, specifically) set about consolidating interwar and wartime research into an integrated research program. Hallmarks of the effort included support for Wiener's incomplete cybernetic studies, the popularization of Shannon's information theory, the establishment of machine translation as an academic field, and grants for launching team-based technical research in fields such as artificial intelligence, linguistics, and psychotherapy. These projects adapted previous decades' programs to emerging Cold War agendas. Funding privileged team-based research that cut across disciplines formed novel institutional alliances and promoted the "worldwide fraternity of scientists" championed by Weaver in the 1930s. Postwar philanthropy for communication research functioned like an epistemic cleaning crew, amalgamating various initiatives underway in decades prior into a durable framework for knowledge production. This agenda deeply shaped the logic of the communication sciences.

For example, the Rockefeller Foundation financed Wiener's adaptation of wartime fire-control research into a mathematical theory of biological feedback. In so doing, it crafted the contours of the "cybernetics moment" with a larger political vision. The Macy conferences offered a forum for speculation on preliminary studies, but the Rockefeller Foundation's funding permitted the development of his work with Rosenblueth into publishable research. The lavish attention given to anecdotes of Wiener's never-built antiaircraft feedback mechanisms reflects the aforementioned myths of wartime invention—an approach that enhances scientific philanthropy's efforts to frame its work as neutral and apolitical. Yet Rockefeller funding for research by Wiener and Rosenblueth allowed for Wiener's single most sustained burst of properly cybernetic research. The decision to fund the project, then, offered a decisive opportunity to support the Macy conferences initiated by Frank, himself a former Rockefeller officer. In their approval of the grant application in 1947, officers at the Rockefeller Foundation praised Wiener and Rosenblueth for exploring how to apply "recent mathematical discoveries to biological and social problems" with the aid of psychiatrists, anthropologists, and sociologists.[79] (In a characteristic instance of synergy between robber baron philanthropies, supplementary funding from the Guggenheim Foundation permitted the polymath Walter Pitts to assist Wiener and Rosenblueth.) Officers expressed the hope that support from the Rockefeller Foundation would establish a "sound experimental basis" for these "at present largely speculative discussions."[80] Whatever results might proceed from a mathematical approach to biology, support for Wiener and Rosenblueth also supported the foundations' longer-term goals for reforming social science.

Funding for Wiener and Rosenblueth also furthered the long-standing goal of promoting cross-institutional and international research, in the form of funded stays at Wiener's institute in the United States and Rosenblueth's institute in Mexico. While their collaboration first developed as an offshoot of wartime research into artillery-fire control, that work had come to an impasse when Rosenblueth left Harvard Medical School for a position in Mexico City. Rockefeller internal documents specified the importance of supporting transnational collaboration over simply bringing the Mexican physician back to an American institution. According to Rockefeller officers in the United States, "much would be lost to the development of Latin American Science if Dr. Rosenblueth and his students, in their enthusiasm for collaboration at any cost, should be driven by a feeling of isolation to seek permanent positions in this country."[81] Foundations often privileged work of this kind, which galvanized knowledge and collaboration across national, institutional, and disciplinary boundaries. The 1947 grant observed that Rosenblueth possessed "a familiarity with mathematics which, though not sufficient for creative work," equipped him for work with the likes of Wiener.[82] The collaboration, officers hoped, would cultivate interdisciplinary methods. In this way, the alliance among the Rockefeller, Macy, and Guggenheim foundations formed an epistemic machinery to circumnavigate academic boundaries.

Foundations' dreams of communication across borders complemented a technocratic tactic—ultimately an antipolitical hypothesis—according to which scientific collaboration would overcome boundaries of race, class, and nation. This conflation of communication and reason conceived of social conflict as mere misunderstanding, and scientific or technological improvements in communication as the natural solution. The funding for Wiener and Rosenblueth, or for the small groups meeting under Macy auspices, was itself a laboratory for overcoming difference through communication. Fremont-Smith, medical director for the Macy Foundation, described a goal of the conferences as "the promotion of meaningful communication between scientific disciplines," which he and the organizers viewed as a real and urgent problem on which the whole of scientific progress rested. "Because of the accelerating rate at which new knowledge is accumulating, and because discoveries in one field so often result from information gained in quite another, channels must be established for the most relevant dissemination of this knowledge."[83] The editors of the Macy conference transcripts, Mead, physicist Heinz von Foerster, and psychologist Hans Lukas Teuber, called the language of communication theory "glimpses of a new lingua franca of science, fragments of a common tongue likely to counteract some of the confusion and complexity of our language."[84] Wiener likewise promoted communication theory as a bulwark against overspecialization and the "comminution of the intellect."[85] The subject of cybernetics was, in the first instance, the scientists themselves, who would establish new foundations for reason through open communication indexed to the mechanisms of relay computers and data transmission rates.

Weaver's New Cultural Analytic

The intellectual labor of Weaver's work promoting information theory, referenced in the introduction to this chapter, is more easily discerned in light of the embedding institutions and agendas. His popular account of information theory as the first step toward "analytical communication studies," published in 1949, distilled decades-long efforts by a sprawling network of institutions and researchers.[86] The crisp empirical clarity of many of Weaver's assertions militated against the claim that they were in any way political. Messages, he claimed, could be modeled as discrete elements distributed in time. The relay of these messages involved senders, channels, and receivers, their work imperiled by noise and governed by laws of statistics and channel capacity. Weaver counted written and oral speech among the most obvious instances of communication of this type but also pointed to music, visual arts, theater, ballet, film, the tracking and targeting of enemy planes, "and in fact all human behavior" as instances of communication covered by Shannon's theory.[87] Where Shannon excluded semantics from information theory, Weaver suggested that in its light "one is now, perhaps for the first time, ready for a real theory of meaning."[88]

Viewed from this perspective of institutional patronage, the conceptual hierarchy between Shannon's and Weaver's theories of communication is not so clear. Shannon penned "A Mathematical Theory of Communication" first, and Weaver subsequently interpreted it. Yet Weaver belonged to and helped cultivate the apparatus in which Shannon's theory and its popularization emerged. Shannon's search for a mathematical theory of communication took shape within this intimate network of scientific patronage that, with particular intensity since the early 1930s, had positioned new research in technical communication as the basis for the fashioning of interdisciplinary scientific alliances. The funding that permitted Shannon to work on the analyzer at MIT, the opportunity to study eugenics, and the wartime cryptography studies that produced the mathematics of information theory developed under the watchful eyes of the administrators of scientific philanthropy. Weaver's popularizing account of information theory crystallized the values of that cultural analytic, celebrating communication, instrumentation, and technical precision as the fundamentals of the coming era of global scientific unity and improved cultural order.

Weaver's efforts from the late 1940s onward to launch machine translation—a topic he promoted in his writings on information theory—offered a complementary consolidation of the new cultural analytic.[89] It distilled the idea that "interpretation," as it was traditionally known, could dispense with literary critics in favor of an impartial and neutral technique for textual analysis. Was this a political or a scientific aim? Did it belong to the concerns of wartime or peacetime? For scientific philanthropy, these oppositions carried little substantive meaning. In a 1947 letter to Wiener on the problem of machine translation, Dr. Weaver suggested that "a most serious problem, for UNESCO and for the constructive and peaceful future of the planet, is the problem of translation, as it unavoidably affects the communication between peoples. . . . I have wondered if it would be un-thinkable to design a computer which would translate."[90] Noting that "the multiplicity of language impedes cultural interchange between the peoples of the earth, and is a serious deterrent to international understanding," he argued that "the world-wide translation problem [could be solved through] . . . the use of electronic computers of great capacity, flexibility, and speed."[91] The foundation's annual report for 1948 explained, "The perfecting of mechanical means of communication—telephone, telegraph, transportation—far outruns our progress in the essential means of communication. These are first of all the construction, utilization and translation of languages. Thus far the most universal language is that of mathematics."[92] Foundation officers believed that bridges built between mathematics and physiology, between Cambridge and Mexico City, and in the mathematical biology of Wiener and Rosenblueth interlinked global reforms of language, politics, and the human sciences.

The postwar propagation of cybernetics, information theory, and game theory in the 1950s spread robber baron philanthropy and its technocratic worldview across disciplines. Former Research Laboratory of Electronics (RLE) director Jerome Wiesner, who would go on to advise the Kennedy White House, remembered the RLE as a hotbed of cross-disciplinary collaboration where neurophysiologists, linguists, economists, and psychologists embraced "concepts of information theory, coding, feedback, prediction, and filtering" as tools for unlocking the human condition. These methods' mathematical formalism and association with engineering promised rigor free from disciplinary bias. Some commentators questioned the suitability of wartime research for peacetime generalization, but few questioned the interwar compact that cemented the vision of engineering as a tool for social order. Spectacular violence between nation-states had the effect of overshadowing more subtle forms of technocratic violence. Debates over the possibility of pacifying wartime research obfuscated how the human sciences had politicized and instrumentalized technical research. Thus the human sciences commandeered wartime research, submitting it to agendas crafted in decades prior by progressive, technocratic intellectuals. International armed conflict, far from militarizing postwar knowledge, veiled how the human sciences and their cultural analytics institutionalized other, more routine and opaque, modes of state force.

Two

Pattern
Recognition

*Data Capture in Colonies,
Clinics, and Suburbs*

Hours of recorded observations, sound recordings and text and photographic series had trained us to be human instruments who could, speedily and accurately, using relatively small amounts of carefully chosen materials, get at a pattern, usable for practical purposes. **Margaret Mead, "From Intuition to Analysis in Communication Research," 1969**

At the previous [Macy] conference I introduced myself by trying to convince those present that zoology, anthropology and psychiatry are really all one and that it is perfectly natural to slide gently from one to the other via an interest in patterns. **Gregory Bateson, Macy *Group Processes* conferences, 1956**

In the 1950s, anthropologist Gregory Bateson and his colleagues in the Palo Alto Group (pictured in figure 2.1) discovered a strange machine in homes across the US suburbs. At the time, US policymakers celebrated the nuclear family as the foundation of national strength and moral rectitude. Right-wingers hailed the nuclear family as a bulwark against state socialism, while liberal and even communist social scientists celebrated it as a resilient agent against brittle fascist mentalities. Both interpretations relied on an industrial-era distinction between adaptive organicism and rigid mechanism, with communism and fascism cast as the political expressions of inhuman political machines. Bateson and his colleagues argued the family was, itself, a cybernetic machine that collapsed distinctions

2.1   *Left to right*: William Fry, John Weakland, Gregory Bateson, and Jay Haley
       of the Palo Alto Group with a tape recorder used to document sessions
       with patients, ca. 1955. A diagram on the blackboard models commu-
       nicative patterns in a three-member family of father (F), mother (M),
       and child (C). The prominence of cameras, recorders, microphones, and
       related media apparatuses in photos of Bateson and his colleagues reflects
       an understanding of the media as constitutive of subjectivity. As Bateson
       once remarked, "It is not communicationally meaningful to ask whether
       the blind man's stick or the scientist's microscope are 'parts' of the man
       who uses them"; human and instrument formed a single observing system.
       Source: Don D. Jackson Archives, University of Louisiana, Monroe, LA.

between machine and organism. Through close analyses of films, photographs, audio recordings, and transcripts of family interactions, they identified informatic patterns structuring everyday domestic life.[1] They viewed
mealtimes, the bathing of babies, and even play among animals as a lively
network of cybernetic encoding and decoding. Successful encoding and
decoding led to a happy, homeostatic family. Contradictory signals, particularly from mothers toward children, produced the "double binds" that
were responsible for schizophrenia. Children afflicted by this and other
mental illnesses tactically blurred levels of communication (e.g., literal and
metaphorical) to produce coherence from these disordered signals. Families, not individuals, were sick, and the person labeled ill, in fact, undertook their communicative amelioration. The Palo Alto Group developed
family therapy as a psychotherapeutic intervention. Their account of the

therapist-as-communication-engineer is perhaps the most successful and enduring application of cybernetics in the human sciences. Their methods continue to thrive under diverse methodological banners including family therapy, brief therapy, and cognitive therapy.[2]

Bateson left the Palo Alto Group in the early 1960s and was thereafter remade, with the aid of the counterculture's leading marketeer Stewart Brand, into a guru of ecological thinking. This status was further enshrined as thinkers including Jürgen Habermas, Gilles Deleuze and Félix Guattari, Jean Baudrillard, and Fredric Jameson adapted his theory of schizophrenia to the description of modern industrial society.[3] This broader acclaim, however, abstracted his theories from the concrete places—colonies, asylums, zoological institutes—that shaped his work. The decidedly technocratic problem-solving underwriting his anthropological and medical investigations disappeared from view. The existential dilemmas of the "schizophrenic" mothers he charged with rendering their children mad as well as the model of ecological relations delineated by caged animals and colonized islanders dissolved into an abstract language of feedback and metacommunications. The result is an impoverished and depoliticized account of what he and the Palo Alto Group contributed to twentieth-century life. Their landmark account of how social and institutional structures produce pathology has become, too often, an idealist language of abstract "codes" and ahistorical "double binds" that circulates without care for the societies and institutions that made that thought possible.

This chapter recovers the contexts that gave rise to the Palo Alto Group, including the partners, problems, and situations that defined its approach to family therapy. It is not a history of Bateson, brilliant theorist though he surely was. Nor is it an account of the great—at times, without parallel—intuitive compassion he and some of his collaborators brought to bear on the study of persons dismissed as "primitive" or "mad" by the dominant culture.[4] The present chapter is, rather, a reconstruction of the generative milieus that shaped elements of the cybernetics and human sciences that persist, albeit in fragmentary forms, today. In particular, this chapter considers how the Palo Alto Group's approach to family therapy reformatted interwar colonial ethnography for the sciences, technologies, and crises convulsing postwar America. Bateson's brilliant and charismatic wife, the anthropologist Margaret Mead, realized much of that theoretical work. In crafting rudiments of a communicative theory of the family through her studies of indigenous kinship systems, she outlined a problematic that would occupy Bateson's work for decades to follow. In the

2.2 Margaret Mead's 1936 visa to enter the Netherlands East Indies for field-work with Gregory Bateson. Source: Margaret Mead Papers and South Pacific Ethnographic Archives, 1838–1996, Manuscript Division, Library of Congress, 194a.

1930s she and Bateson devised methods for the study of Balinese tribes with film, photography, and magnetic tape (figure 2.2). Paired with intricate notation and theoretical reconstruction, their media ethnographies recast indigenous life as symbolic systems shaping activities as diverse as kinship, breastfeeding, and dance. The integrity of these cultural patterns as symbolic systems depended, however, on the unacknowledged role of Dutch colonial violence in isolating and impoverishing indigenous life.[5] And they further relied on the generous funding that came with American philanthropies' hope that illnesses such as schizophrenia (*dementia praecox*) could be explained through experts' technical documentation of people supposedly sheltered from the psychic distortions of modernity. From the 1940s on, Mead and Bateson refined their research in dialogue with mathematicians and engineers, insinuating their presumptions about cultural patterns into broader thinking about the social relevance of computing and informatics. Even more remarkable than family therapy's role in popularizing the cybernetic family was its success recasting colonial kinship as paradigmatic of communication in American families. It was so successful in this endeavor that colonial sources largely disappeared from accounts of cybernetic thought.

In the remarks that follow I reconstruct the itinerary of data capture linking the ethnographies of Mead and Bateson in the 1930s to Palo Alto family therapy of the 1950s and 1960s. As mentioned in chapter 1, Mead and Bateson impacted the formation of the cybernetics group in the 1940s, which partly explains how their colonial research shaped its agendas and methods. This often amounted to an administrative and managerial task. Their specific contribution to the cybernetics of the family emerged from their canny reformatting of colonial media ethnography around the dual crises posed by mental illness and family instability in the United States after World War II. Their myriad contributions to theories of child-rearing in postwar America innovated on a much longer history of using colonized people—as well as soldiers, children, and animals—as material for social theory. To be sure, the Dutch colonial subjects of Bali and Bateson's postwar suburban families around San Francisco were different social formations with distinct histories and ways of life. The violence they experienced, and the political needs they answered, were distinct. This, then, is a history neither of colonial peoples nor of postwar families. Rather, it is a history of shared techniques applied to their management, which formed a network for their analysis by social scientists and policymakers. The entire history of cybernetics is marked, in one way or another, by these kinds of logics—by the processing of problematic populations, texts, diseases, and so on, as models for technocratic reform on a more general level. The analysis I present here therefore traces strategies of datafication as they traveled across colonies, wartime propaganda, and postwar psychotherapies. In the process, the strategies themselves mutated, incorporating new techniques and reorganizing in response to emerging crises. This arc of technical innovation is also an itinerary of media aesthetics in which sources including photography and film shaped ideas of data capture and analysis in the early digital sciences.

Informatics and Aberrancy

According to media theorist Friedrich Kittler, "Since experiments on madmen are usually carried out using the most modern equipment of the moment, they can register with historical precision what in data processing is the state of the art."[6] This dynamic explains the prominence of psychiatrists, psychoanalysts, and psychologists in the early days of communication theory. So, too, it helps explain the requests for assistance and offers of employment that cybernetician Norbert Wiener and information theorist Claude E. Shannon received from social workers and asylums. It drove them to design lavish game-playing machines displayed at theaters, conferences, and on televised broadcasts. There was something irresistibly attractive about the idea that that human thought might succumb to concrete embodiments in new media. In scientific cultures that identify cognition with clear and well-ordered representations, instilling this notion in writing classes and philosophy seminars, it is not so strange to identify the orderly series of film strips and computer programs with human thought. The comparison implies the appealing prospect of technical fixes to the most vexing failures of the human mind.

Family therapy was less an incidental application of cybernetics than an exemplification of the kinds of problems for which it was invented. In the first instance, cybernetics arose as an interdisciplinary project for the human sciences, including technocratic therapies for medical pathologies. The robber baron philanthropies that sponsored cybernetic research seized upon its technocratic possibilities in the human sciences and assembled a stable of researchers to support that agenda. The Josiah Macy Jr. Foundation's focus on medical research and training motivated its early funding for Mead, Bateson, and Wiener in the collaborations later known as the Macy Conferences on Cybernetics. Mead and Bateson, architects of that funding, commanded impressive resumes as experts bringing scientific rigor to bear on cultural problems. As curator of New York's American Museum of Natural History and the most visible anthropologist in US public culture, Mead found lessons for progressive values in the lives of "primitive" peoples. Bateson, the scion of a distinguished family of English scientists, outlined new possibilities for theoretical anthropology with his 1936 book *Naven*. Mead and Bateson hoped their systematizations of social analysis would derive new rigor from the participation of natural scientists. Engineers, in turn, won prestige and influence from social scientists. As early as 1942 the Macy Foundation sponsored meetings on "cerebral inhibition" that were attended by Mead, Bateson, and Arturo Rosenblueth (who presented on his antiaircraft research then underway with Wiener).[7] These dialogues piggybacked on the technocratic dreams of the Progressive Era, shaping the agenda of philanthropies such as the Macy and Rockefeller Foundations.

By the time Wiener published on cybernetics in 1948, his patrons' technocratic concerns with mental health and social order permeated his writing. He looked directly at the moral tragedies of US health policy and made a compelling case that theories drawn from computing could not do much worse. In one particularly bitter remark in his 1948 book *Cybernetics*, he made clear his contempt for the prefrontal lobotomies being routinely administered to long-term patients of mental asylums. The procedures were often justified, he wrote, by "the fact that it makes the custodial care of many patients easier. Let me remark in passing that killing them makes their custodial care still easier."[8] Cybernetics promised to replace the bloody ice pick with elegant studies of signal and noise. He added that the "technique of the psychoanalyst" is entirely consistent with cybernetics.[9] Mathematician John von Neumann deemed these cross-disciplinary applications so promising that he spent the final weeks of his life dictating *The Computer and the Brain* from his deathbed.[10]

Informatic investigation of human aberrancy, however, demanded special situations of enclosure and encoding. Data-driven research into the human condition called for prolonged experimental study, ideally with human subjects available for ongoing surveillance. In keeping with the general thrust of modern experimental science, objects were to be located in discretely ordered time and space, subject to standardized notation by mechanical instruments.[11] The demands of scientific analysis inspired special interest in mental patients, prisoners, children, animals, soldiers, and colonized populations as objects for cybernetic study. These subjects did not have the legal rights or expectations of privacy typically afforded to citizens of Western industrial democracies. Furthermore, exceptional legal and political arrangements seemed to insulate these subjects from the confounding factors of everyday life. Ideologically and infrastructurally, the prison, the reform school, and the colony are often held apart from the relative freedom to circulate that prevails in many liberal democracies. These circumscribed institutions give rise to intensive patterns of communication that also allow privileged regulation and monitoring by authorized regents (guards, social scientists, etc.). On that basis, they promised elementary and uncontaminated data streams from which larger social conclusions could be drawn.

Mead highlighted the usefulness of relatively contained populations for gathering socially relevant data about the human condition. In her landmark 1928 study *Coming of Age in Samoa* she justified the study of indigenous cultures with the claim that Western societies' complexity demanded years of study before "the forces at work within them" could be identified. By contrast, "primitive people without a written language present a much less elaborate problem and a trained student can master the fundamental structure of a primitive society in a few months."[12] This remark encapsulates a concern with the structural and historical conditions, including writing, that permit an economical study of communicative dynamics. Western societies, so the theory goes, have vast systems of scriptural and institutional regulations that complicate analysis. Non-Western cultures without writing seemed to encapsulate their dynamics within concrete practices that permitted total documentation. Mead pointed to children and the insane as analogous groups in Western society suitable for analysis by students of anthropology.[13]

Mental asylums were spheres of protocybernetic analysis in the interwar period when Mead innovated her methods. Consider one such engine for informatic sciences of the mind: New York's Rockland State Hospital (which consulted with Bateson in the 1960s). As early as the 1930s, Rockland had offered data-driven therapy for children with schizophrenia.[14] Among its clinicians in the 1930s was neurophysiologist Warren McCulloch, whose work there on the "mathematico-logical aspects of schizophrenia" directly prefigured his most influential cybernetic studies of the 1940s and 1950s.[15] The hospital also gave rise to the concept of the *cyborg*, an abbreviation of *cybernetic organism*, with its 1960s experimental studies of drugs for medical patients, animals, and future astronauts (figure 2.3).[16] Rockland researchers later turned their attention to the informatic analysis of maladjusted black and Puerto Rican youths from the South Bronx. According to media historian Seth Watter, in this inner-city neighborhood devastated by redlining, poverty, drug abuse, and gang violence, "those who could not adjust, or escape the clutch of law, often found themselves sent on a short drive to Rockland."[17] In the confines of the asylum, therapists collected data on patient behavior and fed it into computers for analysis. In the resulting studies, Watter explains, "patients appear as mere numbers—differences and sums, averages in columns—with neither color nor age nor sex to round out the profile, to bring them before us as once-living persons." In this manner, subjects of legal, behavioral, and psychological aberrancy generated data streams purged of embodied specifics.

Media Ethnography

Adapting theories of cultural patterning advanced by anthropologists Franz Boas and Ruth Benedict, Mead innovated theoretical congruencies between ethnographic observation and statistical theories of information. She sketched a theory of cultural meaning that invited researchers to attend

2.3   Rockland State Hospital researchers who coined the term *cyborg* present "tape from computer which predicted a man's pulse rate from his breath rate." Source: "Man Remade to Live in Space," *Life*, July 11, 1960, 78.

to patterns and systems over intentions and authors. As Benedict put it, the "life history of the individual is first and foremost an accommodation to the patterns and standards traditionally handed down in his community."[18] This suggested a conception of psychopathology as systemic variations of an embedded cultural system. Benedict made her case through the analysis of three tribes: the North American Zuni, whom she characterized as Apollonian, the North American Kwakiutl, whom she labeled Dionysian, and the southwestern Pacific Dobu, whom she described as paranoid. The successful inventorying of native cultures could drive reassessment of the pathologies in Western societies. This analytic accorded epistemic privileges to the material inscriptions captured on recording media, such as the film strip and magnetic tape, that recorded culture as temporal and spatial patterning. The human temptation to identify and interpret could take a backseat to the rigors of media traces available for formal mathematical reduction.

Mead assembled an impressive network of patrons and collaborators for interdisciplinary research around the study of cultural patterns. Her circumspect politics, which overlooked colonial violence and economic inequality to focus on questions of kinship and sexuality, had won her philanthropic patronage from the Social Science Research Council and involved her with Lawrence Frank, who recruited Mead for the interdisciplinary Hanover Seminar of Human Relations.[19] In 1936 Frank moved to the Josiah Macy Jr. Foundation, and his program of funding for small conferences (the so-called Macy conferences) developed in dialogue with Mead and eventually Bateson. The Committee for the Research in Dementia Praecox funded Mead and Bateson's landmark expedition to Bali. According to the two anthropologists, schizophrenic features pervaded Balinese culture but did not seem to cause the same distress as in Western contexts. They argued that the study of schizophrenia among the Balinese would aid its management in the West.[20] In a write-up of their findings not long after the United States entered World War II, they further argued that their ethnographic study would support the coming task of "building a new world . . . richer and more rewarding than any the world has ever seen."[21] While regarding the Balinese as a closed cultural system, the ethnographers viewed themselves as global actors who could leverage knowledge of that system for planetary technocratic reform.

Working in Bali from 1936 to 1939, Mead and Bateson embraced film recording, magnetic tape, and still photography as scriptural apparatuses to anchor cultural research in impartial data patterns. Their 1942 book

Balinese Character: A Photographic Analysis showcased the results of their investigation. It documented what the ethnographers termed a "culturally standardized system of organization of the instincts and emotions" in Bajoeng Gede.[22] Bateson reported treating "the cameras in the field as recording instruments, not as devices for illustrating our theses."[23] Their invocation of "character" subtly revealed the scriptural orientation of the analysis: read as a photographic series, moving bodies functioned as scripted letters, "characters," strung together in symbolic series. Exploiting Bali's reputation as a picturesque destination for Western artists in search of alluring primitivism, Mead and Bateson presented their photographic studies in a format reminiscent of a coffee-table book. Even today, it's often read as a collection of ethnographic snapshots, midway between a museum catalog and album of holiday portraits. A closer reading reveals that the book is an exercise in media surveillance, which presents comprehensive data capture in user-friendly numbered time series (figures 2.4–2.5). Themes present the culture in terms of symbolic exchanges that bind tribespeople into a larger cultural pattern, including "hand postures in arts and trance," "the surface of the body," and "suckling." "By the use of photographs, the wholeness of each piece of behavior can be preserved, while the special cross-referencing desired can be obtained by placing the series of photographs on the same page."[24] *Behaviors* referred not to historical or political actions but, rather, elementary logical forms transmitted across a social body.

The search for impersonal cultural patterns—so much in fashion among liberal New York anthropologists eager to promote social hygiene—carried within it hallmarks of imperial political technologies. For one, Mead and Bateson's reconstructions resembled the obsessive records gathered by Dutch administrators for colonial control. Visual culture theorist Fatimah Tobing Rony has shown how this unacknowledged political background structured the aesthetics of *Balinese Character* and its films. She notes that Mead and Bateson ignored "Balinization," a Dutch policy aimed at reducing Bali to what one historian termed "a living museum."[25] The tribe's semblance of functioning as a self-contained and stable system for visual documentation rested on that history of violent political enclosure. This geographic and political history shaped the very biological organization of the tribe, rendering it uniquely accessible to an ethnographic gaze. The ethnographers credited the tribe's high rates of thyroid disease with encouraging a retardation that produced "schematically simplified" cultural patterns suitable for photographic analysis.[26] These pathologies reflected years of poverty and malnourishment under colonial rule, encoded in

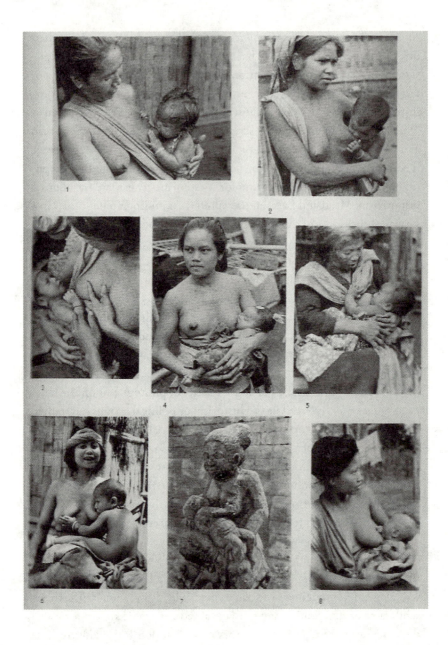

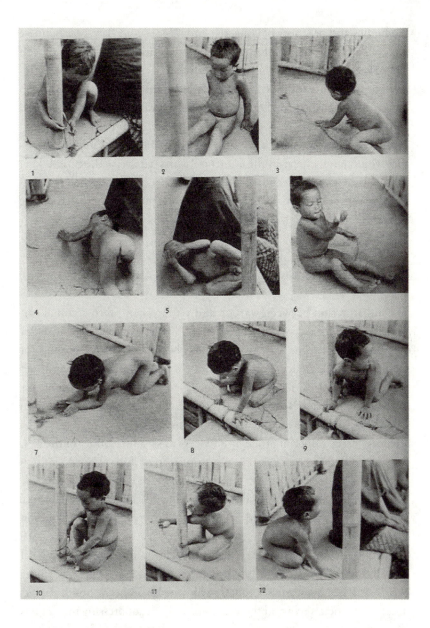

2.4 and 2.5   "Suckling" and a child playing "a bird on a string." These two
series from *Balinese Character* demonstrate the abiding interest of
Bateson and Mead in documenting in mechanical detail serial postures and social interactions, with particular emphases on kinship
and child development. Source: Gregory Bateson and Margaret
Mead, *Balinese Character: A Photographic Analysis* (New York: New
York Academy of Sciences, 1942), 125, 138.

the bodies and practices of collective life.[27] Mead and Bateson's Balinese research assistant I Madé Kaler later said of Dutch rule, "The intention was to keep us from changing and to lead Bali back to the dark ages," adding that such issues never came up with her employers: "Talking was too dangerous, regarding the Dutch. Margaret Mead herself never broached a political discourse."[28] Besides which, Kaler often did the note-taking, capturing testimony with a portable typewriter and choosing the words that would later appear in the anthropologists' analysis. The note-taking praised by Mead as a leap from intuition to analysis embodied colonial forms of objectivity and governance that were expunged from the resulting images of timeless primitive cultures.

Retooling Datafication for the War and Its Aftermath

The indistinction between warfare and welfare, central to colonial surveillance, became a central fact in the mobilization of social scientists for the war effort. In this way, World War II became a proving ground for media-technical sciences cultivated in the Balinese interior. The outlines of a communication science with implications for industrial democracy began to take shape. Bateson wrote in 1942, "It is hardly an exaggeration to say that this war is ideologically about just this—the role of the social sciences."[29] The Office of Strategic Services (OSS), a forerunner of the CIA, dispatched Bateson to Ceylon, India, Burma, and China as a "psychological planner."[30] One such project was operating a radio station to undermine Japanese rule in Burma and Thailand, which included intercepting Japanese propaganda, translating it, and producing counterprogramming. Bateson turned this into an exercise in cybernetic feedback, later explaining, "We listened to the enemy's nonsense, and we professed to be a Japanese official station. Everyday [sic] we simply exaggerated what the enemy was telling people."[31] Later, as an advisor to US occupying forces in the South Pacific, he advised officials to avoid the British mistake of suppressing indigenous communication. He explained this in terms of kinship theories: British attitudes toward colonial subjects, he explained, recapitulated the paternalism of British upper- and middle-class families toward their children. He contrasted this with the dynamic and participatory techniques of American child-rearing, according to which "the American family thus constitutes, in itself, a 'weaning machine'" whose circuits encourage the child to play an active role in self-governing.[32] Nonetheless, he warned that the growing presence of white women in colonies, who generally disapproved of their male counterparts taking indigenous mistresses, had replaced "the more durable and more educative type of relationship [that formerly prevailed] with the native women" with "casual, impermanent and education-al[ly] useless" affairs.[33] The goal of policymakers should be to re-engineer these libidinal relations, strategically linking indigenous groups to a new imperial mother: the US as global superpower.

The interwar apparatus of visual arts, scientific patronage, and technical research, ingeniously crafted by Mead in the 1930s, became a leading instrument of social-scientific strategy in World War II. Together, Mead and Bateson cultivated a new alliance of federal and foundation sponsorship with university and industrial partners. Participating bodies included the Committee for National Morale, which Mead and Bateson joined in 1939 upon returning from Bali; the Council on Intercultural Relations, which they cofounded with Frank in 1942; and the Study of Culture at a Distance project, funded by a $100,000 grant from the Office of Naval Research. This project applied Culture and Personality methods to media-based cultural analysis.[34] In realizing standardized media inscriptions as the foundations of cultural analysis, it sketched new cultural sciences along cybernetic lines. These projects, in turn, enlisted a variety of organizations, including libraries and museums, which put more cultural resources behind the research endeavors.[35] For example, Bateson and former Bauhaus designer Xanti Schawinsky adapted materials from the Balinese expedition into *Bali, Background for War: The Human Problem of Reoccupation*, a 1943 exhibition at the Museum of Modern Art geared toward the forms of cultural understanding conducive to claiming Bali from the Japanese, who had seized it as a wartime conquest (figure 2.6).[36] Serial photographs broke down bodies into discrete series reminiscent of assembly lines.

After World War II, Mead and Bateson reimagined colonial enclosure as a model for global politics. In a 1947 address on how the United Nations could forestall threats from atomic weapons, Bateson argued for far-reaching "social engineering" but warned of the lack of data sufficient to the task.[37] Apart from twenty "small primitive tribes" studied with social problems in mind, only a handful of advanced industrial nations had received thorough anthropological analysis.[38] This differential in ethnographic surveillance reflected asymmetries in political power. Societies subordinated to imperial power and hemmed in by its violence offered better mines for social data. Western political liberties and privileges frustrated comparable analysis. In the coming decades Bateson would redress

2.6 The Museum of Modern Art exhibition *Bali: Background for War, The Human Problem of Reoccupation*. Source: Silvy Chakkalakal, "Ethnographic Art Worlds," *Amerikastudien/American Studies* 63, no. 4 (2018): 507.

that gap by turning to nuclear families whose organizational structures (especially when suburbanized) provided a measure of scientific enclosure, and to mental health crises authorizing observation by external experts such as himself. Mead, for her part, would initiate inventories of the national characters of major powers, from provincial French villages to studies of the Soviet "system."[39] These exercises embodied varied expressions of a project of planetary enclosure that corresponded to the waning of colonial frontiers and the turn toward containment policy to manage threatening global entanglements. This proliferation of enclosures in the Cold War, what historian Paul Edwards termed "closed worlds," carried within itself colonial logics.[40]

Suburban Analysis

Postwar crises over mental health and the family presented key problems against which Mead and Bateson could mobilize a new apparatus of cybernetic expertise. Decades of colonial research into themes such as kinship,

mental health, and parenting found a well-defined and well-funded political priority in the nuclear families settling in postwar American suburbs.
The notion that fascism might have roots in rigid family structures turned
the attention of US-based policymakers and social scientists toward child-
rearing.[41] The therapeutic interest also reflected what historian Ellen Herman characterizes as "an insistent ideology of patriarchal domesticity [that]
simultaneously returned civilian jobs to male veterans and sequestered
women and children in a familial bubble."[42] The family emerged as a key
system for defusing social tensions created by the tasks of reintegrating soldiers into domestic life and withdrawing women from the workplaces they
occupied during men's wartime absence. Technocratic experts and urban
planners turned the suburb into an infrastructure for national strength.

Interest in the family as a stabilizing social system coincided with the
search for new instruments of mental health. Family therapy, founded on
the idea that families, rather than hospitals, could manage psychic tension,
integrated these concerns into a single palliative. During the 1940s the
number of mental patients in the care of the US Veterans Administration
doubled, with 60 percent of the 74,000 patients in its care in 1946 suffering from neuropsychiatric disorders.[43] The National Mental Health Act of
1946 and the founding of the National Institute of Mental Health in 1949
mobilized a national apparatus, as did an expanding network of veterans'
hospitals. Even so, the crisis of the mentally ill seemed to grow inexora-
bly. By the mid-1950s, the number of mental health patients in the United
States had peaked at half a million, filling 50 percent of all hospital beds
nationwide.[44] The demand for mental health care coincided with a tempest
of progressive and reactionary criticism of psychiatry. A spate of journalis-
tic exposés in the late 1940s (such as Albert Maisel's devastating multipage
layout "Bedlam 1946," which appeared in *Life* in May 1946 and in Albert
Deutsch's book *The Shame of the States* in 1947) revealed filthy conditions in many asylums, which often warehoused patients for years on end
(figure 2.7).[45] A flurry of subsequent studies showed that institutionalization itself induced many of the symptoms asylums were intended to treat.[46]
The family, the home, and the suburb appealed as alternative destinations
for treatment.

In concerns surrounding mental health and families Mead and Bateson
found a field ripe for the expertise they had honed in studies of kinship and
schizophrenia in Bali. According to Mead, the American nuclear family
lived in a kind of banishment from tradition, standing in the relative isolation
of single-family homes, bereft of extended networks of care. Every American

NAKEDNESS

AT BYBERRY HOSPITAL NEAR PHILADELPHIA THESE MALE PATIENTS GET NO CLOTHES TO WEAR, LIVE IN FILTH

2.7   *Life* magazine's "Bedlam 1946" showed naked and feeble patients under
       labels such as "filth," "forced labor," and "despair." The images and descrip-
       tions recalled an infamous photo spread dedicated to the German death
       camp Belsen that appeared in *Life* exactly one year earlier. Source: Albert
       Maisel, "Bedlam 1946," *Life*, May 6, 1946, 105.

family organized around its own distinct "code" born "out of the accidents
of honeymoon and parents-in-law, finally beaten into a language that each
[member of the family] understands imperfectly."[47] Balinese tribes, by contrast, had "beautifully precise" code practiced in sonorous harmony across
all members of the tribe. "In a culture like modern America, the child does
not see any such harmonious, repetitive behaviour."[48] The orderliness of a
Balinese tribe, itself an unacknowledged artifact of apolitical ethnography
and colonial enclosure, suggested the outlines of a closed system of communication. In the United States, the isolation of the suburban American family, adapted to the wide-open spaces availed by settler colonialism,
offered a communicative counterpart. A politically specific relation between ethnographers and their colonial subjects became a general feature
in the analysis of US domestic life by one of the nation's leading public
intellectuals.

The breakup of Mead and Bateson's marriage catalyzed cybernetic
studies of family dysfunction. Beginning in 1942 they were intermittently
separated; by 1946 Bateson, returned from his missions for the oss, was

living independently on Staten Island.[49] Bateson soon decamped for a research project on psychiatric communication at San Francisco's Langley Porter Neuropsychiatric Clinic.[50] This set in motion an extraordinary series of media studies from 1949 to 1955 with noted artist Weldon Kees and psychiatrist Jurgen Ruesch.[51] Activities filmed during this period included schizophrenic patients reading *Finnegans Wake*, "mute mongoloids" interacting in group settings, suburban mothers spending time at home with mentally ill children, and seeing-eye dogs undergoing training.[52] Ruesch and Kees undertook a series of landmark studies of "nonverbal communication" that closely emulated *Balinese Character*—albeit with a penchant for humor and irony that seemed to spring from Kees's canny eye (figure 2.8). Film shot by Kees and Bateson at the San Francisco Zoo formed the basis for another educational film that likened caged animals to the human inmates of asylums and prisons, suggesting that animals' tendency to fall into deadening patterns of homeostatic stability provided clues for the treatment of human counterparts. In particular, looking at nonverbal cues that allowed the animals to distinguish between fighting and playing, Bateson worked out the rudiments of the "double-bind" thesis that would dominate the later work of the Palo Alto Group (figure 2.9).[53]

Ruesch and Bateson's coauthored monograph published in 1951, *Communication: The Social Matrix of Psychiatry*, established a cybernetic approach to psychotherapy on the premise that "almost all phenomena included under the traditional heading of psychopathology are disturbances of communication."[54] Echoing Benedict's studies of patterns of culture, they turned away from the expressions of solitary disturbed individuals and toward an analysis of interactions between mentally ill patients and the people around them, including their families and therapists. Ruesch and Bateson turned to Kees to establish techniques for empirically documenting these interactions. A nationally recognized painter and poet, Kees had come to San Francisco to escape the stifling aura of the celebrated New York School of painting with which he was identified. He infused the search for cultural patterns with aesthetic acuity. Kees and his colleagues embraced an empiric of rational analysis purged of the hermeneutic liberties of classical psychoanalysis. Their documentation vibrated with aesthetic excess—a dynamic anticipated in the surreally photogenic quality of the Bali photographs, where visual abundance vouchsafed empirical depth. Kees and company imbued a nascent postwar visual modernism with the authority of unadorned scientific testimony.

2.8   "Individual differences in smoking," one of the many themes by which Jurgen Ruesch and Weldon Kees sought to disclose communication and codification at work in everyday life. Source: Jurgen Ruesch and Weldon Kees, *Nonverbal Communication* (Berkeley: University of California Press, 1956), 61.

2.9   Bateson credited inspiration for the double-bind thesis to his observation of depressed otters living in the San Francisco Zoo, which served as the basis for this educational film shot by Kees. Source: *The Nature of Play: Part 1, River Otters*, directed by Gregory Bateson and Weldon Kees, 1954. Frame enlargement of river otters stimulated to play by the directors. From Bateson EPPI Films, Bateson Collection, Don D. Jackson Archives, University of Louisiana, Monroe, LA.

Cybernetic Families in Postwar America

One of their major works from this period, the educational film *Communication and Interaction in Three Families* (1952), outlined the trio's burgeoning media-aesthetic strategy of analysis.[55] The film, ethnographic in character, attended to subtle patterns of gesture and interaction in three families. Anticipating Mead's suggestions about the unique language of each American family, they approached families around San Francisco living in freestanding, detached homes. They treated each family as an autonomous system of informatic patterning. Eschewing the supposedly impressionistic methods of psychoanalysis and traditional ethnography, they deployed the camera as an apparatus for data capture. A narrator explained, "In this film we have collected, through the camera, data referring to the nonverbal means of communication within the family circle."[56] Unlike in the Balinese work, however, the informatic nature of patterning now came to the fore. Interactions, beliefs, and values were, in their filmic studies, systems of encoding and decoding. "Every expressive movement, action sequence, or word," the narrator explained, "is a message telling the receiver how to interpret other messages. It is in fact a message about communication." Patterns of culture became minute series detectable only with machine assistance. Their identification, in turn, amounted to the discovery of the constitutive features of ethnic and national difference: "small repetitive patterns . . . whose cumulative effect contributes to character formation."

Three Families integrated colonial scrutiny and clinical analysis into a new aesthetic logic of informatic cinema. Indeed, the Langley Porter team viewed cinema and informatics as complementary modes of data capture. Both modes captured data in discrete framed series, permitting new forms of pattern recognition. Ruesch and Kees explained that "few are trained to look steadily and searchingly at the visual world and really see what passes before the eyes," but "putting a 'frame' around a person or group or an object concentrates and emphasizes" observation on empirical reality.[57] Uncovering such patterns was the first step in reforming a prevailing psychiatric gaze dominated by Freudian theories of trauma and depth. "We made a film in '49 at Langley-Porter Clinic," Bateson later recalled, "of the fact that the minor patterns of interchange in a family are the major sources of mental illness. And nobody in '49 could look at that film; the [medical] professionals just could not see it."[58] At stake for Bateson were two competing analytics: psychoanalytic hermeneutics, which sought out traces of traumatic repression in disguised forms (malapropisms, rebus-like dream figures, and bungled actions), and cyber-ethnography, which treated discrete communicative traces as elements in a pseudostochastic communication system.

The Langley team's 1951 film *Hand-Mouth Coordination: Excerpts from the Feeding Routine of a One-Year-Old Boy* (a preliminary study later incorporated into *Three Families*) documented the group's media construction of families as recursive cybernetic systems (figures 2.10–2.13).[59] (It also introduced a blurred relation between system and observer that would dominate Bateson's later work, although the label "second-order cybernetics" tends to exaggerate the novelty of its emphasis on self-reflexivity.[60]) Shot by Bateson and Kees with two portable Bell and Howell 16 mm cameras, the silent film documents in excruciating detail interactions among a mother and her four children, with particular attention paid to the feeding of the toddler.[61] Kees's deft editing turns this ordinary domestic scene into a cybernetic system of gestures and gazes traveling between mother, children, and cameramen. Feeding time at the kitchen table becomes an almost obscene spectacle of orifices as relays: hands summon what mouths swallow, while gazes travel from mother to toddler to brother and back in a constant feedback loop. Drawn into this circuit are the cameramen and their cameras, whose apertures—objects of fascination for the children and of mild distress for a self-conscious mother—become channels for transmitting cybernetic signals.

The cybernetic methodology of the Langley Porter team shifted pathology from the individual body to a widening gyre of familial, architectural, and media relays. The clinic ceded its priority to detached homes and suburban developments. *Three Families* opened with pictures of high rises in an urban center, then cut to a ride along a freeway, before coming to rest at the doorstep of a freestanding single-family home that served as the operative enclosure. This turn toward exurban families, which coincided with an expanding network of suburban clinics, coalesced with the social strategy Raymond Williams memorably termed "mobile privatization."[62] Williams identified it with infrastructures of broadcasting and highways that helped relocate middle-class families from urban centers to privatized suburban homes. The nascent methods of family therapy joined a media-technical ensemble of mobile privatization that included televisions, cul-de-sacs, and picture windows. It allowed regimes of geographic spatialization as well as of economic and ethnic segregation to serve as foundations of family-centered suburban domesticity.

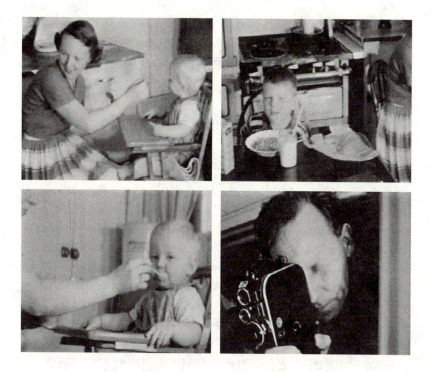

2.10–2.13   *Clockwise from top left*: toddler regards spoon as mother regards son; son regards camera as cameraman (Kees) films son; toddler regards camera as cameraman (Kees) regards toddler; shot by Kees of Bateson filming a wider shot of the whole family. Source: *Hand-Mouth Coordination*, directed by Jurgen Ruesch, Gregory Bateson, and Weldon Kees, ca. 1952.

Yet Bateson's films did not idealize the American family—if anything, they presented it as the composite of artificial codes referenced by Mead above. Recounting an outing in which he took one patient to see his schizophrenogenic mother (a mother whose inconstancy in communication allegedly produced her child's schizophrenia), Bateson grasped suburban, home, and family patterns as a single distorted system. The "nice tract house," he wrote, resembled "a 'model' home—a house which has been furnished by the real estate people in order to sell other houses to the public."[63] Plastic vegetation, china pheasants, and wall-to-wall carpeting adorned the home, lending it an uncanny "exactly as it should be" quality.[64] His judgment seemed to fall not so much on the mother, to whom he ascribed "maternal dominance," but rather on "formal aspects of this traumatic constellation" that permeated the lived environment.[65] Family members, domestic ornamentations, architectural layouts, lawn maintenance, and neighborhood sprawl became part of the communicative codes responsible for schizophrenia. For the patient, this expanse boded peril: "The newspaper boy had tossed the evening paper out in the middle of the lawn, and my patient wanted to get that paper from the middle of that perfect lawn. He came to the edge of the lawn and started to tremble."[66] Architecture and infrastructure entered the technological constitution of the cybernetic family.

Complementing the half hour of calm, domestic scenes that represent the so-called normal families featured in *Three Families* are hours of never-released medical footage of "abnormal" families, shot in the hope of documenting and diagnosing the production of mental illness. There the walls of the family home—presented as neutral and inoffensive in *Three Families*—are disclosed as agents in the often malignant configuration of libidinal economies. In filmed one-on-one sessions between a therapist and the mother of one Palo Alto family (who had sought treatment after seeing a public screening of *Communication and Interaction in Three Families*), the distraught woman complains of the unendurable loneliness she experiences at home, recounting the pain caused by the long absences of her husband and son and declaring, "I'd like to throw rocks through the window . . . [and] tear the walls down."[67] Paired with this agonizing isolation are its unsettling disruptions: she reports distressing phone calls from her husband, visits from an intrusive neighbor, and the overbearing presence of Bateson and his cameraman as incidents that disturb domesticity. Suburban Palo Alto provided a topos for the distribution of this existential dread, and the nuclear family a circuit for its articulation.

The Palo Alto Implementation

In a familiar pattern, it fell to robber baron philanthropies to ensure the translation of experimental communication studies into durable sciences. Faced with precarious funding and tensions with Ruesch, Bateson turned to admirers at the Rockefeller Foundation to help him launch a cybernetic project under his own direction. In a proposal that would form the basis for the Palo Alto Group, Bateson appealed to the foundation for a project at the Palo Alto Veterans Administration Hospital "testing, modifying and amplifying certain hypotheses having to do with . . . paradoxes in communication."[68] When he initially approached the foundation in 1952, officers had mixed feelings about his project. On one hand, Rockefeller president Chester Barnard, formerly of New Jersey Bell Telephone, recognized in it a synthesis of many of his professional and personal interests. In a letter to the president of Stanford University, which Bateson listed as a collaborator on the proposed study, Barnard noted, "Mr. Bateson's plans are of great interest to me personally as they involve original research that cuts across established disciplines especially in theory of communication, psychology, and psychiatry."[69] On the other hand, the officer responsible for reviewing the proposal complained in an internal memorandum that Bateson was not "systemic, analytical or always intelligible. [Indeed,] many of his speculative flights are beyond me."[70] However, the same officer also felt that given the proper framework, Bateson's provocative theses could prove extraordinarily generative. "If he enlists the aid of able young specialists, they will probably bring more discipline to his own thinking," and would in turn implement the theses in a more systemic manner.[71] The strategic goal of Rockefeller financing, then, was not simply to support a promising idea but to convene an interdisciplinary team that would establish more bridges among communication, psychiatry, and anthropology than Bateson himself could.

In June 1952 Bateson received $30,000 from the Rockefeller Foundation to investigate "new problems on the borderline between anthropology, psychiatry and cybernetics."[72] He won follow-up external grants from the Macy Foundation (1954–59), the National Institute of Mental Health (1959–1962), and the Foundation Fund for Research in Psychiatry (1959–62).[73] Bateson hired anthropologist John Weakland, communication analyst Jay Haley, psychiatrist William F. Fry, secretary Margaret Fitzgerald, and Kees to help him gather and analyze data. (Another key member, psychiatrist Don Jackson, did not join the group for another two years.[74]) At the core of their endeavor was a plan to film mental patients in everyday interactions and use techniques derived from cybernetics and information theory to analyze and treat their disorders. Drawing on media analytics forged in New Guinea, honed in the Macy conferences, and polished at Langley Porter, the team embarked on an initiative to examine communicative paradoxes among animals, in humor, and within families. After the 1955 suicide of Kees and the nonrenewal of a Rockefeller Foundation grant, Bateson refocused his efforts around the suburban Palo Alto team, refining means of data capture with media techniques that included audio recorders, film cameras, and one-way mirrors. This collaboration adapted media techniques forged with Ruesch and Kees in the late 1940s and early 1950s into the cybernetic system of inscription, analysis, and therapeutic feedback that came to dominate family therapy and define the Palo Alto Group.

The canonical statement of the Palo Alto Group's work and methods appeared in Bateson's 1956 essay "Toward a Theory of Schizophrenia," which he coauthored with Jackson, Haley, and Weakland.[75] It explained psychopathologies as individuals' constructive and creative response to the paradoxes in communication, with schizophrenia as the privileged example.[76] Later characterized by Haley as "a preliminary report which summarized the common agreement of the research team on the broad outlines of a communication theory of the origin and nature of schizophrenia," the paper argued that mental illness sprang from communicative contradictions in the family, which the authors termed "double binds."[77] They cited the case of a schizophrenic patient whose mother verbally demanded affection from her son but physically withdrew when he embraced her. Bateson and his team claimed that schizophrenic symptoms, such as confusion between literal and metaphorical levels of communication, provided a tactical and rational resolution of double binds. Where gesture and speech conflicted, or affective and discursive levels of communication clashed, the schizophrenic produced a mash-up of interpretive frames that reconciled competing levels of meaning. From this analytical perspective, the best method for curing schizophrenia was to engineer a change in the familial patterns of communication responsible for its genesis.

From the mid-1950s through the early 1960s the Palo Alto Group devised an array of media-technical systems to analyze and modify patterns of communication. Bateson, wary of experimental laboratory research, chafed at the techniques, precipitating his departure from the project in 1962.[78] As was the case in Bali years earlier, he was ambivalent about wielding media technology as a constructive force in social formations. For years to come he would largely abstract his theoretical projects from the political infrastructures that gave rise to them.[79] His colleagues in the Palo Alto Group showed no similar compunctions; if anything, they were quite chuffed to show off their technical innovations. One notable setup was the structured interview, which they deployed as an assemblage of cameras, microphones, architectural designs, scripts, and games to aid therapists in "scanning for patterns" in families.[80] Most often this system was used in interviews of families with a member who had been labeled as schizophrenic. In its simplest form, the structured interview consisted of standardized questions posed by a family therapist. These questions organized (or "structured") topics of discussion, the preferred sequence of family members' responses, and the presence or priority of particular family members (or the therapist) during the conversation. The paradigmatic structured interview distributed the family in a horseshoe shape around a table with family members facing inward and the therapist and movie camera positioned at the open end of the horseshoe. Although occasionally performed at home, structured interviews often took place in a purpose-built room that included a table, chairs, a fireplace, and a one-way mirror that concealed the camera and observing therapists (figure 2.14).[81]

The focus of the structured interview on binding patterns in family interactions often entailed a recourse to game theory. One iteration asked family members, individually and in isolation, what they believed the main problem in the family to be. Afterward, they were brought together, informed of unnamed discrepancies among their individual accounts, and challenged to reach a consensus about the true nature of the problem. A therapist observed as the family negotiated, and a camera documented from behind the one-way mirror in the hope of revealing double-binding processes in action. "This task," a member of the Palo Alto Group explained, "is an adaptation of the game-theoretical model of the Prisoners' Dilemma where, as is known, direct communication is made impossible, a decision involving all concerned has to be reached and the decision is dependent upon the amount of trust each partner is prepared to invest in the others."[82] In these and other tasks, it was "not so much the content of their final decision which has been found to be revealing . . . but whether or not a decision is reached within the time limits, and the manner in which it was accomplished."[83] Therapists recorded, transcribed, and minutely analyzed interactions structured by game theory to manifest interpsychic double binding.

The media setup systematized therapist and family as a single pattern of cybernetic communication. Frequently therapists called for patients' attention for purposes that included policing quarrels or compelling the faithful performance of tasks, but particularly emphasizing their mutual inscription by media; therapists, no less than family members, were subject to the impartial record of the film and magnetic tape.[84] This last point encouraged family members to focus on immanent networks of interactions in which the camera positioned the therapist as a node rather than as an external authority. This regulation of participants' performance redounded on the therapists themselves, who found in celluloid and magnetic tape

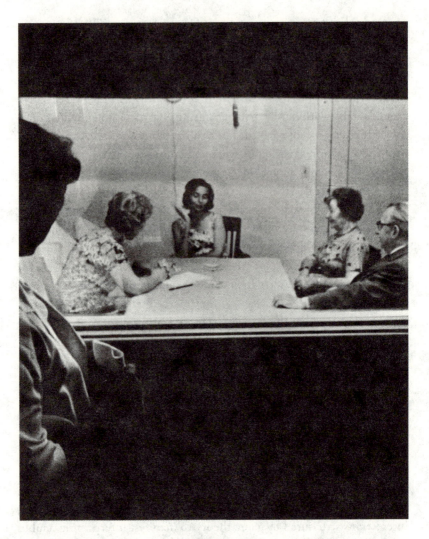

2.14 Photograph taken from behind the one-way mirror of an interview room at the Mental Research Institute, Palo Alto, California. Originally captioned "Schizophrenia patient Ida Friedberg (end of table) and parents in a Palo Alto therapy session." Source: Milton Silverman and Margaret Silverman, "Psychiatry inside the Family Circle," *Saturday Evening Post*, July 28, 1962.

an objective record that short-circuited the temptation of psychoanalytic interpretation. "The ultimate verification of typical family patterns," group member Haley explained, "would seem to be possible only if the family is placed in some experimental situation where the responses can be recorded by some other means than the quick eye of an investigator."[85] Family members and therapists became part of a single informatic pattern regulated by media technology.

The structuring power of the Palo Alto Group's apparatus reached its apex in an experimental setup (once again relying on game-theoretical models) for tracking patterns of family cooperation and competition devised with Alex Bavelas of MIT's Group Networks Laboratory (figures 2.15–2.16). Members of a family were asked to play a game of alliance-building. For example, each member of a three-person family (e.g., child, mother, and father) took a different place in a cubicle setup. The walls of the cubby obstructed family members' lines of sight, preventing visual signaling among them. Communication thus took place by means of switches. Each partition had two buttons, individually labeled to designate the person in the adjacent partition (mother, father, or child). When family members from adjacent cubbies simultaneously pressed the buttons designating each other, they formed a coalition and were awarded a point. Pairs were exclusive—only two people could be in a coalition at a time—and to win the game a family member had to shift back and forth strategically between coalitions with different family members. For example, if a mother formed a coalition only with a son and vice versa (i.e., never engaging the father), then no one would win the game because the mother and son would have the same number of points and the father none. An "event recorder" in the adjacent "control room" kept a master record of all interactions for later analysis and comparison across families. Apparatuses such as these facilitated formalist analyses of family interactions. More decisively, they performatively reconstituted the family into a rule-bound communication system governed by elementary operations of relay and exchange.

In contrast to classical psychoanalysis, which remained rooted in Western ideas of the self, the Palo Alto analytic found its roots in colonial ethnography focused on collective ways of life. Apparatuses and experiments deployed by the Palo Alto Group stripped away conversational semantics to reveal elementary patterns of interaction such as those an ethnographer with limited knowledge of a tribe might observe. With cybernetics, information theory, and game theory "updating" these methods for postwar science, a rich interiority of patients' lived knowledge was transformed

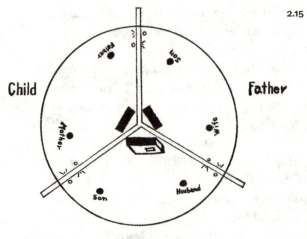

2.15 Diagram of an apparatus designed to standardize and quantify patterns of interaction in family members, aimed at comparing the patterns of "normal" families with those of families with a child labeled schizophrenic, ca. 1962. Source: Jay Haley, "Family Experiments: A New Type of Experimentation," *Family Process* 1, no. 2 (1962): 283.

2.16 Photo of a five-person setup of the "scanning for patterns" apparatus developed by Bavelas and reportedly tested with US soldiers. Source: Alex Bavelas, "Group Size, Interaction, and Structural Environment," *Group Processes: Transactions of the Fourth Conference*, ed. Bertram Schaffner (New York: Corlies, Macy, 1959), 134.

into patterned signals circulating a collective. Therapists drew out data series and fine-tuned their patterns. Where psychoanalysis would have read repetitions, gestures, stutters, and cross talk in terms of the depths and obfuscations of the psyche, family therapy treated these elements as depersonalized codes governing an informatic network. Psychotherapeutic depth and repression thus gave way to cybernetic pattern, captured in a material surface of celluloid and magnetic tape. Indeed, staff members sometimes divided up family members' spoken lines and acted them out with one another: if mental illnesses resided in patterns of interaction, then even "normal" subjects—so the logic went—might slip into psychically aberrant roles.[86]

These methodologies had a peculiarly centrifugal effect on mental ill-ness. No longer inscribed in the flesh, mental disturbances became am-bulatory, traveling from body to body, without a center, the property of ongoing interactions rather than internal states. This shift of locale cor-responded to modifications in documentation and treatment. Recording media gave illness a durable existence in material traces outside the body; illness no longer resided in tissues of gray matter but rather in informatic patterns. By means of recording, these traces could be sped up, slowed down, cut into pieces, redistributed across distances long and short, and circu-lated for analysis via professional journals. In time, these results could be returned to the patient—sometimes even directly, through an encoun-ter with words and images played back to them, for instance. But more often, therapists synthesized these elements and applied their lessons to the family, thereby completing the circuit of communication with the ther-apist becoming the final segment of the feedback loop that reconstituted the family as a cybernetic system.

By 1962, the year Bateson left the Palo Alto Group, its methods had gained early recognition. Efforts to reproduce its members' work were tak-ing root in experimental clinics across the United States and later in Europe and South America. In 1967, Bateson's former colleagues and students in Palo Alto established the Brief Therapy Center, which adapted the lessons of family therapy for the treatment of patients identified as suffering from a dearth of "time, money, intelligence, persistence, and verbal sophistica-tion."[87] For proponents, family therapy evoked a larger epistemic reform of the arts and sciences. One practitioner boldly told a journalist for the *New Yorker* that family therapy would soon eclipse psychiatry because "it is a therapy that belongs to our century, while individual therapy belongs to the nineteenth century."[88] What "Pinter is to theatre and ecology is to natural science," he explained, family therapy would be to psychiatry.[89] But brief therapy overshadowed family therapy and Pinter (the jury is still out on ecology). As therapists across the United States embraced these meth-ods in the 1980s, one critic referred to the new methods as "therapy in the age of Reaganomics."[90] The skepticism toward society encouraged by burgeoning neoliberalism found its complement in therapies that treated families and individuals as free-floating systems. The permanent exile of Balinese tribes, conceived as outside history and politics and excluded from Western social contracts, now became the social exile of Thatcher-ite families. No society; no state; individuals and families, aided by entre-preneurial technicians, became pivot points of social order. In lieu of the vast machinery of cameras, transcriptions, mirrors, and microphones, brief therapy returned to the simple cybernetic system of therapist, notebook, and patient. In doctrinaire form, it afforded no more than ten or twelve sessions of intensive work in which to identify and adjust aberrations in communication. By taking funds formerly directed to institutionalization and psychoanalysis and earmarking them for brief therapy and its spinoffs (e.g., cognitive behavioral therapy), HMOs and state healthcare systems became the principal apparatuses for disseminating stripped-down cybernetics to the salaried masses.[91] Today, anyone who has gone through ten or twelve sessions of cognitive-behavioral therapy may count themselves as an extended member of the cybernetic family.

Three

Poeticizing Cybernetics

An Informatic Infrastructure
for Structural Linguistics

Structural linguistics shifts from the study of conscious linguistic phenomena to study of their unconscious infrastructure. **Claude Lévi-Strauss, "Structural Analysis in Linguistics and Anthropology," 1945**

On November 16, 1944, the Linguistic Circle of New York gathered at the Manhattan auditorium of the American Telephone and Telegraph (AT&T) to witness a performance of the Voder, a synthetic speaking device that built up sounds, words, and sentences from a phonetic keyboard (see figures 3.1 and 3.2). Russian-born linguist Roman Jakobson had founded the group in 1943 as a successor to the celebrated Prague Linguistic Circle, where he had spent much of the 1920s and 1930s developing structural linguistics from a synthesis of the work of Swiss linguist Ferdinand de Saussure with diverse insights from fields including the theory of relativity, cubism, and Russian formalism. With the arrival of German troops in Prague in 1938, the circle collapsed and Jakobson fled, eventually arriving in New York in 1941 as a stateless person. Now he sought to establish a foothold for structural linguistics in the New World, and he believed recent research in telephony would assist him in that effort. Fellow member of the New York circle, French ethnographer Claude Lévi-Strauss, whom Jakobson had personally initiated into structural methods, later explained that the Voder, along with the loosely related information theory of AT&T

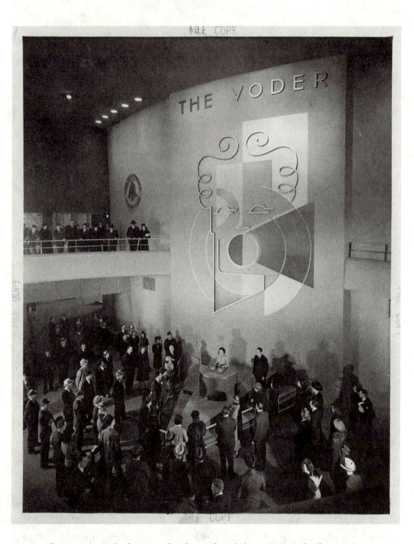

3.1 "American Telephone and Telegraph Exhibit—The Voder," Press Department, Bell Telephone Exhibit, New York World's Fair, Queens, New York, 1930–40. Source: Image 1652487, Manuscripts and Archives Division, New York Public Library, http://digitalcollections.nypl.org/items /5e66b3e8–9002-d471-e040-e00a180654d7.

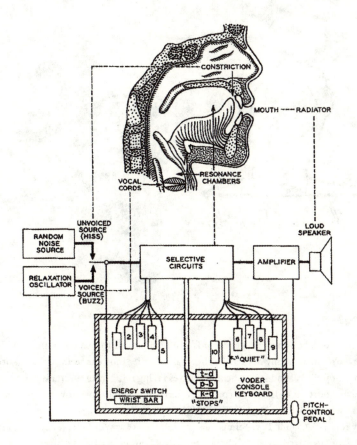

3.2 Reconceptualization of the vocal apparatus as a discrete series of instruments modeled on the Voder, from a pamphlet produced by the Bell Labs System for exhibitions of the Voder. Source: Roman Jakobson Papers, box 6, folder 74, Roman Jakobson Papers, MIT Distinctive Collections, Cambridge, MA.

research engineer Claude E. Shannon, instantiated the "main principles of interpretation" of Saussurean linguistics"[1] (figures 3.3 and 3.4). Among these principles were the following: "that communication among people is based on the combination of ordered elements; that, in each language, the possibilities of combination are governed by a group of compatibilities and incompatibilities; and lastly, that the freedom of choice in language, as defined within the limits of these rules, is subject, in time, to certain probabilities."[2] Over the coming decades, veterans of the New York circle, among them Jakobson, Lévi-Strauss, and Thomas Sebeok, would transform AT&T

3.3 and 3.4

Portraits of Jakobson from FBI records and Lévi-Strauss from a US committee for refugee scholars. Many of their intellectual activities took shape amid the negotiations of political and scientific imperatives facing refugee intellectuals. Sources: Undated portrait of Jakobson (probably derived from passport or visa paperwork), file 105–706, 1A-1, FBI Boston Field Office, Special Agent James Newpher, received May 10, 1956. Claude Lévi-Strauss, ca. 1941. Image 1687584, Emergency Committee in Aid of Displaced Foreign Scholars records, Manuscripts and Archives Division, New York Public Library.

researchers' models into cornerstones of structural theories sweeping linguistics, literary studies, anthropology, and semiotics.

The Linguistic Circle of New York's trip to AT&T offers a glimpse into how midcentury scientific infrastructures incorporated European human sciences into emerging logics of US communication science.[3] It highlights how strategic US research conglomerations during World War II shaped a new theoretical order among the natural and human sciences in subsequent decades. Jakobson and his colleagues' 1944 field trip to AT&T belatedly convened disparate wings of communication research rallied by the United States against wartime enemies. On their way into the auditorium that day, members streamed past an emblem of that project: a granite relief of the United States, its coasts linked by a transcontinental wire, adorned with the inscription "Service to the Nation at Peace and War" (figure 3.5). The Voder's source device, the Vocoder, had recently figured

3.5 "Service to Nation in Peace and War," installed in the Manhattan head-quarters of the Bell Telephone System, championed the dual roles of the Bell system in securing national unity through internal improvements (represented by a map of the east and west coasts of the United States connected by telegraphy) and military support in World War I. Chester Beach, granite, 1928, photographed by Morris Rosenfeld. Source: Smithsonian American Art Museum and Renwick Gallery.

in cryptographic technologies permitting secure conversations between Winston Churchill and Franklin Delano Roosevelt (technologies from which Shannon's information theory would soon develop).[4] This same instrumental amalgamation had incorporated the scientific knowledge of the circle. Jakobson and Lévi-Strauss taught in the United States as part of a strategic program to ally European scientists with American science. Other members consulted for the Office of Strategic Services (oss, predecessor to the CIA) and the US War Department. Their colleagues and interlocutors around the city, among them anthropologists Margaret Mead and Gregory Bateson, likewise consulted on wartime communication research. These arrangements located structural linguistics within an emerging theoretical program that grasped cybernetics, information theory, game theory, and computing as parts of a single program of communication science.

This chapter examines the strategic refashioning by Jakobson of structural linguistics according to the fraught circumstances of revolutionary Moscow, interwar Prague, wartime New York, and Cold War Cambridge, Massachusetts. His incorporation of structural methods into information theory and cybernetics, according to the anti-Soviet programs of Cold War science, figured prominently in his reinvention of structuralism. In one sense, that effort recalls the previous two chapters: a charismatic, globe-trotting researcher aligned himself with the technocratic priorities of US funders, notably the Rockefeller Foundation, with an eye to emerging political priorities. In the process, he re-established a distinguished theoretical program to permit the cybernetic management of a disruptive other. For Jakobson, this other was not the social consequences of industrialization, nor colonized and suburbanized kinship systems, but rather Russian speakers. With the aid of MIT colleagues and the Rockefeller Foundation, Jakobson spent the 1950s modeling a cybernetic enclosure of the Russian language, crafting theories and methods that became the commonplaces of semiotic and literary theory of subsequent decades. Yet Jakobson's deep background in poetics, elaborated in Moscow and Prague amid purges, pogroms, and genocides, imparted distinct political priorities to his work. In this respect, Jakobson's engagement with cybernetic topics often evoked his erstwhile years as a Russian futurist, fusing poetics and technics against unspeakable violence.[5] As we will see in chapters 4 and 5, that backdrop would have far-reaching consequences for the critical reception of cybernetic structuralism across the humanities.

The Politics of Structuralism

Cybernetics and structuralism grappled with grim political realities according to complementary notions of communication and enclosure. Both belonged to a family of communication theories aimed at manufacturing political order. It testified to the United States' effort as a burgeoning superpower to grapple, by means of communication engineering, with global conflict. Its emphasis on human action enclosed by impersonal technical and political systems led Paul Edwards to liken it to a "closed world." Highlighting its close ties to military conflict, fellow American historian Peter Galison called it an "ontology of the enemy," in which friend and foe faced off from within radically opposed communicative networks.[6] The deeper roots of these systems lay in a settler-colonial mentality of resolving individual difference through technical communication. One public service advertisement for AT&T in 1920 depicted Christopher Columbus and his crew disembarking on a beach with the caption "Our Many-Tongued Ancestors" (figure 3.6).[7] Accompanying text clarified, "A confusion of tongues makes for a confusion of ideas. Everything which goes toward the up-building and maintenance of a one language people makes for national strength and national progress. It is in such service that the Bell Telephone has played so vital a part." Native Americans and enslaved Africans did not figure into this proposal. Communication theory elaborated the myth of a self-made American people, erasing ethnic difference to produce an ostensibly unmarked space of unchecked communication.[8] This agenda fit within Bell Telephone's oft-touted commitment to "one policy, one system, universal service" that accommodated a diverse nation to one protocol. The systematizing horizon of US communication theory took for granted this subordination of difference to an integrated technological and political authority.

The structuralists—mostly Europeans, many of whom spent significant periods of their lives as refugees, exiles, and expatriates—hailed from traditions that viewed already existing cultures as enclosing, at times totalizing, systems. As linguists, ethnographers, and psychologists, their fields inventoried cultural patterns in service to an imperial state. Educated at conservative universities with premodern roots, in societies forged through centuries of religious and ethnic strife, they viewed language and culture as systems structuring individual action. The systemic assignments of culture could not be built anew around improved engineering schemes. Rather, they bound the individual on a course fixed according to

Our Many-Tongued Ancestors

Born of the diverse nations of the earth, Americans appreciate, now more than ever before, the necessity for national unity; one flag, one purpose, one form of patriotic understanding.

A confusion of tongues makes for a confusion of ideas and principles. Everything which goes toward the up-building and maintenance of a one language people makes for national strength and national progress.

It is in such service that the Bell Telephone has played so vital a part. Its wires reach every corner of the country, making

intimate, personal speech between all kinds of people a matter of constant occurrence.

But the telephone is no interpreter. If its far reaching wires are to be effective, those who use them must speak the same language. The telephone best serves those who have become one with us in speech.

Yet uniformity of language is not enough from those who would gain the greatest good from the telephone, neither is financial support enough; for complete service makes essential true co-operation on the part of every subscriber.

AMERICAN TELEPHONE AND TELEGRAPH COMPANY
AND ASSOCIATED COMPANIES

One Policy   *One System*   *Universal Service*

3.6

"Our Many-Tongued Ancestors" was one of dozens of advertisements in the early decades of the twentieth century presenting AT&T and the Bell Telephone System infrastructure as vehicles of cultural unification and homogenization. Images of ships in the background, presumably of the *Niña*, *Pinta*, and *Santa Maria*, faded due to the age of the newsprint, remain visible. Source: American Telephone and Telegraph Company, "Our Many-Tongued Ancestors," *Survey* 43, no. 19 (March 6, 1920): 715.

impersonal collective relations. In the work of three structuralists we find three such conceptions. Following longtime service to the French Republic in support of its effort to project linguistic hegemony across its dialectically varied population, Saussure famously likened language to a game of chess. No mere instance of free individual expression, it rather implied agonistic, rule-bound constraints as the condition of articulation. One did not simply speak but selected from among a field of possibilities prefigured by formal constraints and anticipated threats. Lévi-Strauss's more melancholic oeuvre reconstructed the structural logic of cultures decimated by colonial expansion. In writing of *mythologiques* (mytho-logics), he proposed structured lawlike operations to indigenous cultures available for deductive reconstruction in the wake of their colonial decimation. French psychoanalyst Jacques Lacan saw in criminality and madness impersonal, collective rules acted out by individual bodies.[9] As such, the incorporation of structural methods into US infrastructures was not so much a fall into politics as a reckoning with its changing conditions.

In sixty-odd years as a linguist, Jakobson grappled in the most existential terms with the fraught role his field played in political struggle. Endowed with a genius for synthesis and compelled by political threat to flee one university after another, during a long career he finally outstripped Saussure himself in terms of promulgating structural analysis. In some respects, the course of Jakobson's career follows the course of twentieth-century elite Western science. From storied crucibles in Central and Eastern Europe it took flight across Western Europe, eventually assuming a Cold War perch at Harvard University that gave it the resources to reach out again to the rest of the world. Jakobson founded the Moscow Linguistic Circle as an undergraduate in the 1910s, amid the tumult of linguistics, poetics, and aesthetics on the eve of the Russian Revolution. A fellow traveler of the Russian futurist poets, he found in language a medium invested with the force and threat of modern life. While doing fieldwork in 1914, gathering folklore in the countryside with a colleague from the circle, he narrowly escaped murder at the hands of locals who mistook him for a German spy.[10] When the revolution swept Moscow, the nascent socialist republic enlisted him in a project to bolster territorial claims by defining linguistic frontiers with Ukraine. He accepted the assignment in exchange for passports permitting his parents to flee the country.[11] Jakobson followed in 1920, at the age of twenty-three, when he joined a Red Cross mission destined for Prague.[12] Many less fortunate peers perished in the first decade of the Soviet republic, their language and arts ill-suited to the new doctrines. One such friend, formalist literary critic Viktor Borisovich Shklovsky, whom Jakobson had hidden in premises that served as the circle's library and salon, ultimately took refuge in Finland. In 1922 he wrote to painter Ilya Repin to ask for help returning safely. The latter replied, "You write, asking that I certify that you are not a Bolshevik, and you write the letter in the Bolshevik orthography. How can I possibly defend you?"[13]

Against this backdrop of political-linguistic terror, the structuralist enterprise took shape under the auspices of a newly launched Prague Linguistic Circle. Jakobson cofounded the group with the refugee Russian prince Nikolai Trubetzkoy. Borrowing ideas from cubism, physics, and Russian formalism, they systematized the sketchy lectures of Saussure into an organized doctrine of language as a system that produced meaning through differential relations among its terms. Each language has a fixed ensemble of terms defined not by their positive identities but rather by the "distinctive features" that set them apart, which could be analyzed in terms of binary alternatives such as vocalic/nonvocalic, consonantal/nonconsonantal, compact/diffuse, and tense/lax. In this analysis was a theory of communication that armored language against the violent eruptions that drove Jakobson and Trubetzkoy from their homelands. Systematicity and identity were relative, rather than absolute, qualities. A language, a dialect, or a single sound distinguished itself through dynamic and ongoing adaptation to adjacent terms. In one of the most noted works of the Prague Circle, "The Principles of Historical Phonology," Jakobson posited that the disruptive forces of historical change within a language, its diachronic axis, were in fact a mechanism for animating lively and stable synchronic dynamics.[14] Historical change—the shedding of distinctive features, the emergence of new words—no longer deformed a language but instead marked the way the synchronic axis maintained change. Historical breaks became an instrument of securing, rather than suspending, stability in the communication system. As rendered by Jakobson, a language system counteracted potential disruptions by communicating transformations up and down its entire linguistic chain. This sustained the telos of *langue* toward an orderly equilibrium that never devolved to the ultimate stable state—death—but instead kept the entire system in a state of mutual adaptation among its parts that resulted in dynamic stability.

Jakobson's structural methods, particularly as they blended with the ethnographically inflected studies of Lévi-Strauss and linguist Émile Benveniste, seemed to elicit the essential features of a language precisely as it faced liquidation. As violence swept across a land, reducing a culture to fragments, structural methods extracted "differential features" that identified the quintessence without retaining the original material. Jakobson unfolded these themes in "The Generation That Squandered Its Poets" (1930), written to honor his suicided friend, the Russian Futurist poet Vladimir Majakovskij. In it, Jakobson opposed lively, dynamic poetics to the ossifying effects of rigid, historicized language. He describes himself and Majakovskij as part of a "lost generation" that arrived at the Russian Revolution, "not rigidified, still capable of adapting to experience and change, still capable of a dynamic rather than static understanding of our situation."[15] This phrasing recapitulates the Prague Circle's rejection of philology in favor of structuralist methods that treated language as changeable, generative, and oriented toward future transformations. "All we had were compelling songs of the future; and suddenly these songs were transformed by the dynamics of the day into a historico-literary fact."[16] As French philosopher Jacques Derrida later argued, structuralism possessed a "catastrophic consciousness, simultaneously destroyed and destructive."[17] Faced with cataclysmic threat, it took the measure of a culture, identifying essential systems and features and discarding excess baggage with the brutal efficiency summoned by the threat of imminent destruction.

Destructive force followed structuralism. As cataclysmic violence had struck down the Moscow Circle, the outbreak of World War II would crush its successor. Not long after Trubetzkoy assumed a professorship in Vienna, he perished of a heart attack apparently brought on by the stress of his perilous anti-Hitler agitation following Austria's annexation. Soon German troops were in Prague, and Jakobson, warned of personal danger by a friend, decamped first for Denmark, then Norway and Sweden.[18] Before leaving he reduced his vast collection of personal papers to "nine pails of ashes" for fear that the Germans would track down his correspondents. His wife recalled the ensuing years as a time of dislocations, temporary storage, documentation, filing, stamping, signing, and code-switching: "Temporary apartments, clothes in suitcases, boxes in storages, losses during transportation from one country to another, looking for new apartments, visas, places on boats and trains, switching from one language to another, from one environment to another, and people, people, people of all countries, characters, professions, destinies."[19] Their arrival in New York in 1941 offered a partial refuge from persecution. Malevolent political attention intermittently focused on the Jewish Russian intellectual known for his cosmopolitan ties reaching through the Iron Curtain. It took intervention from his onetime colleague at Columbia University turned US president, Dwight Eisenhower, to apparently squelch a public summons from the House Un-American Activities Committee (though declassified files suggest private testimony ensued).[20] In his ardent support for Slavic studies in the 1950s and 1960s, one sees an effort to rescue a literary culture crushed by political enemies tactically hitched to a knowledge production meant to curry patronage from a never entirely trustworthy US political establishment.

Structural Linguistics in a Cybernetic Key

Ultimately, it was US political strategy that won Jakobson refuge in the United States and set his research on a path of informatic investigation. As war swept Europe, the Rockefeller Foundation mobilized to bring threatened European intellectuals under the umbrella of US wartime science. The École libre des hautes études (ELHE), hosted by the New School for Social Research in New York City, carved out a special niche for francophone intellectuals in exile. In 1941, Alvin Johnson, the president of the New School, approached the foundation with a proposal to establish a university for French scholars imperiled by the Vichy regime. He sought to establish in New York what Officer Tracy Kittredge characterized as "collaborative research programs related to world problems of economic, political, legal and social character."[21] Johnson brokered an agreement whereby he would select candidates for fellowships to teach at the ELHE while officers at the Rockefeller Foundation would approve or reject funding for those candidates (Lacan, for example, was summarily rejected for a fellowship in 1942).[22] Support would be limited to a few years, after which time professors were to return to Europe or assume normal professorships in American universities.

Upon securing a professorship at the ELHE, Jakobson began retooling structural linguistics for US communication research. Already in the *Course in General Linguistics*, Saussure had characterized the organs for speech production as a "vocal apparatus" (see figures 3.7 and 3.2) and promoted the use of film to develop a scientific technique for studying the articulation of sounds.[23] However, he had balanced these instrumental overtures with a sharp delineation between the apparatus for the production or study of speech and the material of speech itself. As Saussure put it in one lecture, the "vocal organs are as external to language (*la langue*) as the electrical apparatus which is used to tap out Morse code is external to that code."[24] In lectures at the ELHE, Jakobson instead presented media technology as a precondition of structuralist investigation. He praised "development in telephonic communication, radio and sound in Europe and, above all, in America" with accustoming listeners to attending to sound independent of its speaker.[25] Jakobson likewise credited X-ray photography and related technologies with validating many of Saussure's theses, suggesting that technical limits to recording and graphing speech formed a leading obstacle to continued advances.[26] By creating durable inscriptions of ephemeral sound, these instruments presented speech as a physical object appropriate for study in its own right.[27]

In the late 1940s, as Europeans were streaming back to their home countries, the Rockefeller Foundation embarked on a new program of cybernetic research loosely tailored to an emerging Cold War threat. Jakobson sought to remain in the United States and offered himself as a resource for the emerging agenda. In correspondence with the foundation, he touted "the

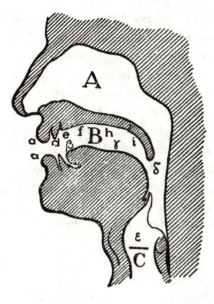

3.7

Diagram of "the Vocal Apparatus."
Source: Ferdinand de Saussure,
Course in General Linguistics, ed.
Charles Bally and Albert Sechehaye,
trans. Wade Baskin (New York: Phil-
osophical Library, 1959), 41.

amazing progress of acoustics, particularly of the revealing results of the
Bell Telephone Laboratories' research, and in view of the growing X-Ray
and sound film techniques" as a model for future linguistic research. He
predicted that "startling results could be swiftly achieved if the sound patterns of various languages were systematically subjected to acoustic analysis
jointly guided by linguists, researchers in physical acoustics, and experts in
problems of sound perception."[28] In 1949 Harvard University appointed
him to the professorship in Slavic studies formerly held by Norbert Wiener's father, Leo Wiener.[29] This prestigious position brought him greater
access to a network of researchers in communication engineering across
Cambridge. In a letter to the Rockefeller Foundation, he proposed that
N. Wiener, MIT's Research Laboratory of Electronics (RLE), and Harvard's
Psycho-Acoustic Laboratory collaborate on a new project to apply information theory to the study of Russian phonemes. Touting a compact integration of political, epistemological, and technological interests, he wrote
to officers at the foundation that the exhaustive analysis of present-day
standard Russian, founded on the synthesis of American and Western
European science, would offer a scientific advance that Soviet science,
"terrorized by doctrinary purges and paralyzed by a narrow-minded un-
productive official bias, is unable to accomplish."[30]

The Rockefeller Foundation embraced Jakobson's mixture of scientific
universalism and partisan politics. In 1950, he received a $50,000 five-year

grant under a new foundation humanities program, Language, Logic, and Symbolism, which was followed up by an additional $80,000 of funding before the end of the decade.[31] In a report to the foundation's trustees, President Barnard maintained, "such an analysis may facilitate the application to living languages of the mathematical theory of communication worked out by Mr. Claude E. Shannon and Mr. Warren Weaver."[32] This fit with a larger effort to orient Rockefeller funding toward anti-Soviet research. In his dispatch to trustees, Barnard asserted that given the world's present circumstances, "the ivory tower attitude would be as unreasonable as the iron curtain attitude."[33] The battle against both would be waged not on the barricades or in the trenches but in the laboratories, with exiled linguists, refugee literary critics, and erudite folklorists trained by information theory and armed with spectrograms.

Toward a Cybernetic Archive of Russian

Jakobson's Rockefeller-funded project, titled "Description and Analysis of Contemporary Standard Russian" (1950–58), gave rise to a new kind of structural linguistics. Take Jakobson's 1956 *Fundamentals of Language*, coauthored with MIT linguist Morris Halle. In *Fundamentals*, the coauthors adapted terms from information theory, cybernetics, and game theory to the structural analysis of language. Linguistic articulation became an exercise in informational distribution. Much as Shannon used probabilistic sequences to predict series of words, phrases, and sentences, Jakobson and Halle described phonemes as probabilistically encoded and decoded series. By way of example, they painted the scene of a New York party guest introduced to one "Mr. Ditter." For the guest, recognizing the name involves rapidly cycling through possible combinations of sounds. The guest searches for the correct identification of the utterance, considering terms such as *bitter*, *dotter*, *digger*, and *ditty*. In lieu of positively identifying a name, they explained, recognition became a matter of differentiation among an ensemble of possible selections. Jakobson and Halle characterized these distinctions as "information-bearing elements."[34] The presentation suggested ever accumulating series of ascending and descending binary selections, all available for machine analysis (figure 3.8). *Ditter*, for example, comprises four sequential units: /d/+/í/+/t/+/ə/, each of which admits various combinatoric reductions and juxtapositions.[35] This analysis of oral communication as vying combinatorics—both hearer and speaker governed by the probabil-

	k	k,	g	g,	x	c	ʃ	ʒ	t	t,	d	d,	s	s,	z	z,	š	n	n,	p	p,
VOCALIC	−	−	−	−	−	−	−	−	−	−	−	−	−	−	−	−	−	−	−	−	−
CONSONANTAL	+	+	+	+	+	+	+	+	+	+	+	+	+	+	+	+	+	+	+	+	+
COMPACT	+	+	+	+	+	+	+	−	−	−	−	−	−	−	−	−	−	−	−	−	−
DIFFUSE	0	0	0	0	0	0	0	0	0	0	0	0	0	0	0	0	0	0	0	0	0
GRAVE	+	+	+	+	+	−	−	−	−	−	−	−	−	−	−	−	−	−	−	+	+
NASAL	0	0	0	0	0	0	0	0	−	−	−	−	−	−	−	−	−	+	+	−	−
CONTINUANT	−	−	−	−	+	−	+	+	−	−	−	−	+	+	+	+	−	0	0	−	−
VOICED	−	−	+	+	0	0	−	+	−	−	+	+	−	−	+	+	−	−	−	−	−
SHARP	−	+	−	+	0	0	0	0	−	+	−	+	−	+	−	+	0	−	+	−	+
STRIDENT	0	0	0	0	0	0	0	0	−	−	−	−	0	0	0	0	+	0	0	0	0
STRESSED	0	0	0	0	0	0	0	0	0	0	0	0	0	0	0	0	0	0	0	0	0

	b	b,	f	f,	v	v,	m	m,	ʲu	u	ʲo	e	ʲi	i	ʲa	a	r	r,	l	l,	j
VOCALIC	−	−	−	−	−	−	−	−	+	+	+	+	+	+	+	+	+	+	+	+	−
CONSONANTAL	+	+	+	+	+	+	+	+	−	−	−	−	−	−	−	−	+	+	+	+	−
COMPACT	−	−	−	−	−	−	−	−	−	−	−	−	−	−	+	+	0	0	0	0	0
DIFFUSE	0	0	0	0	0	0	0	0	+	+	−	−	+	+	0	0	0	0	0	0	0
GRAVE	+	+	+	+	+	+	+	+	+	+	+	−	−	−	0	0	0	0	0	0	0
NASAL	−	−	−	−	−	−	+	+	0	0	0	0	0	0	0	0	0	0	0	0	0
CONTINUANT	−	−	+	+	+	+	−	−	0	0	0	0	0	0	0	0	−	−	+	+	0
VOICED	+	+	−	−	+	+	−	−	0	0	0	0	0	0	0	0	0	0	0	0	0
SHARP	−	+	−	+	−	+	−	+	0	0	0	0	0	0	0	0	−	+	−	+	0
STRIDENT	0	0	0	0	0	0	0	0	0	0	0	0	0	0	0	0	0	0	0	0	0
STRESSED	0	0	0	0	0	0	0	0	+	−	0	0	+	−	+	−	0	0	0	0	0

3.8 Graph devised by information theorist Colin Cherry and linguists Morris Halle and Roman Jakobson to reduce Russian phonemes to binary oppositions (coded as plus or minus) permitting measure of their value in bits according to Claude E. Shannon's theory of information. Source: Colin E. Cherry, Morris Halle, and Roman Jakobson, "Toward the Logical Description of Languages in Their Phonemic Aspect," *Language* 29, no. 1 (March 1953): 38.

ity of a fixed ensemble and its possible chains—adapted Saussurean ideas of a speech community to a nascent algorithmic account of language.

Fundamentals of Language formed part of a far-reaching account of language as a cybernetic system centered around Jakobson's studies of Russian. His works from this period, including *Preliminaries to Speech Analysis* (Jakobson, Fant, and Halle, 1951), "Toward the Logical Description of Languages in Their Phonemic Aspect" (Cherry, Halle, and Jakobson, 1953), and "Shifters, Verbal Categories, and the Russian Verb" (1957) crafted a cybernetic framework for structural analysis. His approach to linguistic acts as the processing, storage, and transmission of data simultaneously rewove the fabrics of language and the sciences. The ephemeral synchronic dimension of speech—both the keystone to Saussure's critique of philological approaches to language and uniquely resistant to experimental inquiry or empirical demonstration—received orderly, economic expression according to the methods of information theory. Language became probabilistic and combinatoric, ordered on principles that echoed Saussure but following the directions of cybernetics. The operations of algorithms became one and the same as the operations of speech.

From Jakobson's research a new kind of "theory" for the human sciences took shape. It dissolved differences between engineering and culture, suggesting a single framework of encoding, decoding, and feedback ruled over both. In a 1952 presentation to a gathering of anthropologists and linguists attended by Lévi-Strauss, Jakobson remarked, "We have involuntarily discussed in terms specifically [of communication engineers], of encoders, decoders, redundancy, etc."[36] Anxious not only to borrow from the harder sciences but also to give back to them, Jakobson added, "Communication theory seems to me a good school for present-day linguists, just as structural linguistics is a useful school for communication engineering."[37] In parallel, Jakobson developed a book series promoting communication theory in the human sciences with communication engineer Leo L. Beranek and linguist William Locke, both of the RLE. Lévi-Strauss, Lacan, and W. V. Quine were among those who promised contributions.[38] He further accepted an appointment to a professorship at the RLE, an editorial position on the interdisciplinary journal *Information and Control*, and a seat on the steering committee of MIT's Center for Communication Sciences.[39] MIT president Julius Stratton expressed the institute's confidence in and hopes for Jakobson in a 1957 letter, writing, "We share fully your conviction that the problems of communication and language will occupy a place of increasing importance in all modern science."[40]

Where analysts like Mead and Bateson had found in cybernetics a shell for theorizing colonial and schizophrenic subjectivity, Jakobson found a system for characterizing a new threatening other: the Soviet speaker of Russian. If these theories availed the structural linguistics of a burgeoning infrastructure of communication engineering, so too did they position structural linguistics as a carrier of US scientific agendas. Jakobson recruited colleagues around the world to contribute to his work, impressing the interest of cybernetic analytics on theorists from Julia Kristeva to Umberto Eco. This work, moreover, formed part of a larger knowledge infrastructure targeting the Soviet state. A Rockefeller-funded word count at Wayne State University adapted Russian speech to computational analytics. Punched cards coding Russian words aligned structural, cybernetic, and political concerns (figure 3.9). It coded terms for computational sorting in terms of noun, adjective, pronoun, composite word, masculine, feminine, transitive, intransitive, reflexive, root word, and prefix. The result was a portrait of language as digital transformations, codable to permit algorithmic extrapolations capable of simulating more complex interrelations. Meanwhile, a Carnegie Corporation–funded project at Harvard's Russian Research Center gathered cybernetic anthropologists including Clyde Kluckhohn for the systems-minded analysis *How the Soviet System Works*.[41] Faced with an intractable enemy, these initiatives suggested that

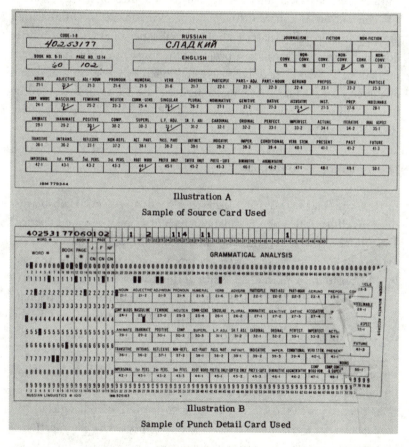

Illustration A

Sample of Source Card Used

Illustration B

Sample of Punch Detail Card Used

3.9 Punched card used in the Russian word-count project at Wayne State University. These exercises sought to accommodate language and culture to the analytical possibilities of early computing. Source: Harry Hirsch Josselson, *The Russian Word Count and Frequency Analysis of Grammatical Categories of Standard Literary Russian* (Detroit, MI: Wayne State University Press, 1953), 37.

cybernetics, allied with fields such as linguistics and anthropology, could produce a stable inventory of cultural and political differences.

In Jakobson's case, however, this vast cybernetic capture also carried in it a refugee's ambition to restore lost worlds. Within the proposed exhaustive description of the Russian language, realized with the aid of engineers and spectrograms, dwelled a utopian urge. It stood alongside *Mimesis*, German critic Erich Auerbach's 1946 tribute to a European literary canon, written while exiled in Istanbul and remote from his library. It likewise paralleled Austrian novelist Stefan Zweig's 1942 memoir *The World from Yesterday*, an encomium to a disappeared Habsburg Empire. These works gathered up cultures crushed by war and political change. They memorialized languages and conventions not merely driven out of their habitats by authoritarian purges but irreparably shattered, indispensable personas and sensibilities not having survived wars. These projects functioned as archives against annihilation. The database of language envisioned by Jakobson indexed an imagined system of language, made necessary by the annihilation of the literary and aesthetic world he once knew.

The Language of Database

In 1958, the final year of the Rockefeller-funded project, Jakobson distilled his efforts into a programmatic statement, "Linguistics and Poetics." This work both accomplished and abolished ambitions for the cybernetic mastery of language according to a quintessentially structuralist analysis. In it he posited the mutual necessity of linguistics, poetics, and literary criticism, suggesting the possibility of their joint accommodation to Shannon's schematic account of communication (figures 3.10 and 3.11). In this respect, it fulfilled the longstanding aims of MIT colleagues and robber baron philanthropies to reorient the humanities toward exact, quantifying, empirical, and rule-governed theoretical analysis. Jakobson's claims rested, however, on the idea that linguistics, in its mathematizing ambitions, needed poetics to specify the ultimate distribution of language. Not only speech but mathematics itself must turn to poetics for illumination of its distinctive feature. The pairing of "linguistics" and "poetics" thus stood in for a more general relation of communication theory to poetic revelation, in which each term became known through its relation to the other. In a theme that would become central to later structuralist and poststructuralist criticism, a science of poetic critique implied the necessity of a poetic critique of science.

The situation of the lecture partly illuminates its ambitions. It served as the closing lecture to a "Style in Language" conference hosted by Sebeok and funded by the Social Science Research Council (SSRC). The conference brought together communication-minded researchers such as cognitive scientist George A. Miller and literary critic I. A. Richards to introduce theoretical rigor into the fuzzy concept of "style." The centerpiece of the talk, Jakobson's account of poetics modeled on Shannon's schematic diagram of communication, inventoried the poetic function as including the *addresser* (sender) of the *message* (conflation of message and signal) sent to an *addressee* (receiver), the *contact* (channel), and transmitting or representing *code* (code). In embracing these terms, Jakobson echoed the speakers that came before him, many of whom likewise invoked information theory (figure 3.12). Here, the diagrammatic strategies of communication engineering imposed an orderly set of distributions and series on the unruly multiplicity of language performances. Having recast Saussurean *langue* and *parole* as informational *code* and *message*, Jakobson's theory presented speakers as perpetually consulting economic codes to convey messages in rule-compliant terms. In certain instances, style might be expressed according to subcodes (corresponding to dialects and subcommunities), but these remained loosely indexed to the constraints of the "over-all code" that was the "unity of language" for its speakers.[42] As Shannon had asserted that "selection" governed the construction of Markov chains in communication, Jakobson now posited that selection and combination structured the "speech chain."

With Jakobson's analysis in place, linguists could join engineers in the laboratory, rub elbows, share ideas, and take part in wrenching language from the amorphous domain where Saussure had left it and reinstall it within a modern scientific program instead. Through this refashioning of linguistic acts as a techno-economic matrix of production, Jakobson provided mechanisms for conjoining poetics with a theoretical system ascendant from psychology to economics. He realized in the field of language arts the kind of interdisciplinary, mechanistic account of communication, positioned to pacify political passions with reasoned scientific technique, that launched the mission of "scientific philanthropy" in the early decades of the century. By aligning the refined conceptual systems of interwar Central European thought with the communicationism of midcentury American

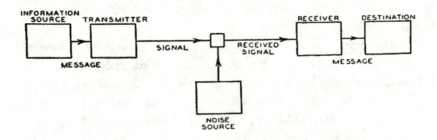

3.10 Claude E. Shannon's schematic diagram of a communication system. Source: Claude E. Shannon, "The Mathematical Theory of Communication," in *The Mathematical Theory of Communication* (Urbana: University of Illinois Press, 1949), 5.

CONTEXT

ADDRESSER   MESSAGE   ADDRESSEE

CONTACT

CODE

3.11 Roman Jakobson's schematic diagram of constitutive aspects of speech. Source: Roman Jakobson, "Closing Statements: Linguistics and Poetics" (1958), in *Style and Language*, ed. Thomas A. Sebeok (Cambridge, MA: MIT Press, 1960), 353.

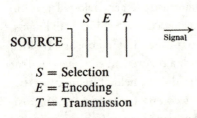

S = Selection R = Reception
E = Encoding D = Decoding
T = Transmission DV = Development

3.12 I. A. Richards's gloss on the schematic diagram of communication. Source: I. A. Richards, "Variant Readings and Misreading" (1958), in *Style in Language*, ed. Thomas A. Sebeok (Cambridge, MA: MIT Press, 1960), 242.

science, Jakobson envisioned his own particular axis of global fraternity, closely tied to forces of Western capitalist production. (He colorfully illustrated this technoscientific fraternity when he entered a Harvard lecture hall one day to discover that the Russian economist Vassily Leontieff, who had just finished using the room, had left his celebrated account of economic input and output functions on the blackboard. As Jakobson's students moved to erase the board he declared, "Stop, I will lecture with this scheme." As he explained, "The problems of output and input in linguistics and economics are exactly the same."[43])

Jakobson's cybernetic analysis performed a structuralist twist: where other scholars invoked communication theory to stabilize stylistic utterances, Jakobson submitted theory itself to stylistic analysis. Here, in Bloomington, Indiana, was the voice of Russian futurism and Prague linguistics, a dynamic unfolding of technics and poetics, in an oscillation that was more structuring than structure. As noted above, at the level of rhetoric, the title "Linguistics and Poetics" amounts to a structuralist pairing whereby a juxtaposition of elements permits distinctive features to emerge through contrast. Jakobson observed, "Whenever the addresser and/or the addressee need to check up whether they use the same code, speech is focused on the CODE," as when a speaker refers to the discourse itself to ask, "Do you know what I mean?"[44] This reflexive operation, which he terms *metalanguage*, invites comparison with poetics because it calls attention to the selection and combination of terms. "Poetry and metalanguage, however, are in diametrical opposition to each other: in metalanguage the sequence is used to build an equation, whereas in poetry the equation is used to build a sequence."[45] Metalanguage makes equations out of discourse whereas poetics makes discourse out of equations. Where one stabilizes a discourse to assign ratios in communication, the poetic function in fact reflects on these ratios to generate new discourses. Where metalanguage rises above speech to grasp the order of discourse, as in the form of an equation, poetics sinks deep within the language, animating the intrinsic mathematical relations among terms to provoke a synthetic rediscovery of the possibilities of the code.

Jakobson's poetic function was not merely an *application* but also an *analysis* of communication theory. Poetic language, like other modes of discourse, "strives to build an equation" out of verbal terms, but in an essentially indeterminate manner that countervails the tendency of metalanguage to stabilize meaning and clarify codes. "Let us repeat with Empson," Jakobson intoned, "the machinations of ambiguity are among the very roots

of poetry," calibrated to render the "referential function . . . ambiguous."[46]
This valorization of poetics displayed Jakobson's fidelity to the circles of
Moscow and Prague. Decades earlier among the futurist poets, Jakobson's
dream of complementarity between scientific and poetic functions had
taken shape. His decades-long program to identify structuralism with the
relativistic molecular relations unearthed by atomic physics and cubism
rested on understanding them as complementary analytical schemes. To assign information theory the status of a master science capable of clarifying
these principles under the sign of *bit, feedback*, or *code* would have merely
substituted what he once ruefully termed "historico-literary fact" for a
new cyber-communicative fact.[47] Only through the ongoing transvalua-
tion among terms could structuralist insight make a difference. Jakobson's
efforts to grant equal epistemological stature to poetics won few followers
among engineers and communication scientists. It was rather students of
poetics, particularly in modernizing postwar France, who first heeded his
call. It is to that discursive migration that we now turn.

Four

Theory for Administrators

The Ambivalent Technocracy
of Claude Lévi-Strauss

As entropy increases, the universe, and all closed systems in the universe, tend naturally to deteriorate and lose their distinctiveness, to move from the least to the most probable state, from a state of organization and differentiation in which distinctions and forms exist, to a state of chaos and sameness. . . . [But] there are local enclaves whose direction seems opposed to that of the universe at large and in which there is a limited and temporary tendency for organization to increase. Life finds its home in some of these enclaves. It is with this point of view at its core that the new science of Cybernetics began its development. **Norbert Wiener, *The Human Use of Human Beings*, 1950**

The contrast between a 1930s exhibition and a 1960s laboratory captures how computing changed conceptualizations of data in the human sciences. In 1937 the married ethnologist duo of Claude and Dina Lévi-Strauss (née Dina Dreyfus) presented *Indians of Mato Grosso*, an exhibition in Paris of artifacts from an expedition to Brazil (figure 4.1).[1] As with relics of a saint, the spiritual meaning of indigenous artifacts inhered in an incalculable materiality. This emphasis on indexical traces—photography, pottery, and so on—gave way in the 1950s to a new account of ethnography as the study of informatic systems such as those explored by cybernetics and computing. Claude Lévi-Strauss's Laboratory of Social Anthropology (LSA), founded in 1960, teased out the logical structures of primitive life with codes, punched cards, mathematics, and structural analysis.

Its towering Human Relations Area Files (HRAF) compiled ethnographic findings worldwide into a searchable database of two million index cards (figure 4.2). Primitive life itself, Lévi-Strauss argued, mirrored patterns of logical computation. Writing in 1963 in the *Mercure de France*, Dina charged that her erstwhile husband's informatic turn of mind reduced primitives to little more than "well-conditioned machines."[2] Of his pre-occupation with how primitives and moderns logically ordered everyday life, she noted that neurotics often showed similar passion for inventorying their closets. She asked, archly, "Should we mistake this classifying frenzy for some kind of speculative power?"

This chapter recovers the deep ambivalence of Lévi-Strauss's cybernetic turn of mind, marked by a conflicted relationship to liberal technocracy. If Dreyfus's commentary evokes the incredulity that shadowed the ascent of Lévi-Strauss's structuralism and its cybernetic ideals in postwar France, it hardly outstrips his own complex attitude toward these methods. Like the cybernetician Norbert Wiener, Lévi-Strauss identified communication theory with a modern world exploited by ceaseless and abstract exploitation. What Lévi-Strauss described as the "consolidation of social anthropology, economics, and linguistics into one great field, that of communication" concerned not merely unification but also destruction.[3] Singular and concrete realities collapsed into the numerical abstractions prized by Western science. Thus, borrowing a term from cybernetics and information theory, he suggested rechristening anthropology *entropology*: a science of decay and de-differentiation set in motion by the ceaseless relay of words, practices, and goods globally. Rather than merely oppose communication, however, Lévi-Strauss sought to rally its techniques in a new project of cultural analytics, whose rise I outline in this chapter.

Technocracy at Cross-Purposes in French and US Human Sciences

The imperial roots of linguistics and ethnography deeply marked their reception of cybernetics and its promulgation in postwar France.[4] Histories of French technocracy lavish great attention on engineers and functionaries, often neglecting technocracy's role in nascent human sciences.[5] Through inventories of cultural and linguistic practices, nineteenth-century ethnographers and linguists helped the French empire define and shape the lives of its subjects. Their long-standing emphasis on the processing of cultural

MUSÉUM NATIONAL D'HISTOIRE NATURELLE

MUSÉE DE L'HOMME

UN INDIEN BORORO

INDIENS DU
MATTO-GROSSO
(Mission Claude et Dina Lévi-Strauss)
Novembre 1935-Mars 1936

GUIDE-CATALOGUE DE L'EXPOSITION
ORGANISÉE A LA GALERIE DE LA
"GAZETTE DES BEAUX-ARTS" ET DE "BEAUX-ARTS"
21 Janvier - 3 Février 1937

4.1

Cover of the catalog for the exhibition of photographs and artifacts collected by Claude and Dina Lévi-Strauss. Source: *Indiens du Matto-Grosso (Mission Claude et Dina Lévi-Strauss), Novembre 1935–Mars 1936, Guide-catalogue de l'exposition organisée à la galerie de la "Gazette des Beaux-Arts" et de "Beaux-Arts," 21 Janvier–3 Février 1937* (Paris: Musée de l'Homme, 1937).

4.2

Claude Lévi-Strauss amid the vast Human Relations Area Files. Source: Marion Abélès, "Le fichier des Human Relations Area Files," *La lettre du Collège de France*, Hors-série 2 (2008): 66–67.

data with technical precision served the infrastructural, educational, and administrative ambitions of their patron. For all that, technocracy did not easily translate into the sphere of neutral expertise it promised. Servants of technocratic government did not necessarily subscribe to the ideals of nonpolitical or impartial expertise. Their practical experiences often confounded such premises, leading to theories that militated against dreams of technical stabilization.[6] By the same measure, across time and place, definitions of technical neutrality varied tremendously, such that different political regimes advanced conflicting notions of neutrality. All these factors would come to shape the reception of cybernetic ideas in postwar France.

Consider linguist Ferdinand de Saussure, the progenitor of structuralism, whose work suggested that the rigorous analysis of empirical data revealed the cultural systems that produce distinct communities. As a researcher at the École pratique des hautes études (EPHE) from 1881 to 1891, Saussure contributed to the French republican mission of outlining the structural regularities of its linguistic populations. From this project emerged the rudiments of a structural analysis, "shot through with the political and social interests of the French state" and mediated by the scriptural methods of the "phonetics laboratory."[7] Even so, his embrace of differential terms over positive identity implied a critique of French republicanism. No homogeneous citizenry, equal before the state, could really be developed from a cultural order premised around difference. Decades later, writers such as critic Roland Barthes and philosopher Jacques Derrida elaborated these implications into meditations on how ethnic and racial difference figured within language.

Technocratic tensions shaped French ethnography, notably the work of protostructuralist Marcell Mauss. Born in 1872, Mauss had studied religious sciences at the EPHE in the 1890s and had taught there since the early 1900s. Writing to a colleague at the École française d'Extrême-Orient in Hanoi, Mauss asserted frankly, "Colonial policy may be the area in which the adage 'knowledge is power' is best confirmed. By respecting and using beliefs and customs, modifying the economic and technological system only with caution, not opposing anything directly, and using everything, [administrators] could arrive at humane, easy, and productive colonial practices."[8] When he founded the EPHE's Department of Ethnology in 1925, he promised it would stand "at the disposal of colonial governments and protectorates for any information concerning expeditions (French or foreign), the study of the indigenous races, the conservation and study of monuments and collections, or the study of social facts."[9] Faced with civil servants who preferred to pacify colonies with canons and had little compunction about incinerating the records of ancient civilizations, Mauss's proposals exemplified a spirit of liberal technocracy. Against brute violence, he volunteered technical expertise as an antidote to social tensions. In practice, this amounted to political enclosure and inclusion, including a circumscription of French republican values.

Rockefeller charities' early efforts to intervene in French social science reveal the distinct political values of competing technocracies. French technocracy emphasized centralized groups of experts who were credentialed and underwritten by the state. Saussure and Mauss turned to the state and its instruments for funding, aligning their aims with its patronage. The systems builders charged with producing national unity—be it through railways or telegraphy—more often hailed from industry. Structuralism, with its emphasis on collective over individual initiative, encoded these values.[10] The American conceptualization of technocracy, by contrast, showed greater interest in private industry and initiative. Considering these differing approaches to technocratic social science, it is not surprising that the Rockefeller philanthropies made slow progress in their outreach to French intellectuals in the 1920s and 1930s. Their model of privately funded research and ad hoc institutes would prove ill-adapted to the entrenched forms of French scientific research. Their general preference for establishing institutes and research topics independent from existing institutional structures rejected the dominant logic of state-planned social science that defined French expertise of the period.

The Rockefeller Foundation's vexed relationship with Mauss illustrates the competing technocratic values of US and French social science. In 1917, an early incarnation of the Rockefeller Foundation's social sciences division, the Laura Spelman Rockefeller Memorial (LSRM), established an office for reforming European social sciences. It soon cultivated relations with Mauss, the favored nephew of Émile Durkheim. The LSRM paid for his travel to the United States in 1926 so that he could learn about the social-scientific methods being developed there and give lectures on French ethnography.[11] This was part of the Rockefeller Foundation's interwar program for "cross-fertilization" between national scientific communities. Among his various activities, Mauss gave a lecture at Harvard and at the University of Chicago on "The Unity of the Human Sciences and Their Mutual Relationship: Anthropology, Psychology, Social Science." With the support of the LSRM, the University of Chicago at that time had become one of the world's leading centers for the integration of technocratic, theoretical, and practical social science. It exemplified the promise that technocratic expertise could guide social-scientific initiative. Social scientists' influence in policy circles impressed Mauss.[12] He hailed the achievement of the "great" American people that had placed "its entire social system, its entire demographic composition, as well as its destiny and its full individuality under the jurisdiction of a practical reason finally enlightened by science and, in any case, rationally managed by scientists and by the people themselves."[13]

In 1929, at the invitation of the Rockefeller philanthropies, Mauss applied for funding to establish a new center for social science in Paris.[14] He proposed the establishment of a faculty of social science at the École pratique des hautes études—the so-called Sixième Section, since it would have been the sixth faculty housed at the EPHE. Mauss argued such a center would gather the scattered activities of French social science under one roof, fostering a form of unity befitting their object of study. "The unity of the social sciences," he wrote in his application, "will be demonstrated only when all teachers and all students, whatever their area of specialization in that vast field, are obliged to meet, and do meet, in a place where the material means for work and contact have been expanded."[15] The officers of the Rockefeller Foundation, however, balked at his proposal, invoking a number of empirical and methodological concerns. They complained his plans were too vast, too vague, too abstract, and unlikely to make serious contributions to social control.[16] They instead offered a lavish subvention of $350,000 to Charles Rist, an economist who also served on the governing board of the Bank of France, to establish an institute of economic and social science. The foundation designated additional funds for training students and smaller grants for more modest initiatives to "familiarize the younger elements at the university with the methods of observation and the work necessary to solve economic, sociological, and political problems" and to develop "true methods for social control."[17]

Given its enthusiasm in courting Mauss, the foundation's rejection of his proposal is striking. In internal memoranda, Tracy Kittredge, the assistant director of the social sciences division at the Rockefeller Foundation's Paris office, complained bitterly of the theoretical and speculative methods that prevailed within French social science, which in his view remained detached from concrete social problems.[18] Rockefeller-funded initiatives routinely delimited specific problems among specific populations—literacy among rural African Americans, appreciation of "American" traditions at universities, the promotion of Basic English at select Chinese universities, the cultivation of political science in London—and convened committees to promote these ventures. This strategy of intervention corresponded with a rational and technical style of reasoning that studied phenomena in parts (or more precisely, that viewed the individual as more foundational than collectives or relations) with the aim of manipulating and reforming individual elements in a society. Mauss's search for "the unity of the social sciences," by contrast, expressed a holistic conception of society inflected by French republican ideals. It rejected the underlying logic—individualistic, atomizing, and oriented toward private initiative—that guided Rockefeller-funded social science of the 1920s and 1930s.

More to the point, Mauss viewed the individualizing and calculated pragmatism promoted by the Rockefeller Foundation with suspicion. Not only opposed to American technocratic reasoning, Mauss also cast doubt on the technocratic ideals of interwar social democracy in Europe. Consider his 1925 *Essai sur le don*, published in English as *The Gift: The Form and Reason for Exchange in Archaic Societies*. Ostensibly a study of practices of exchange, it concluded in a stirring rebuke of the modern liberal state's myth of a neutral adjudication of social interests.[19] Written amid the great debt crisis that crippled Germany after World War I, Mauss found in the reciprocal practices governing indigenous potlatch evidence of the impossibility of a purely technical solution to the problem of competing social aims.[20] He showed that gift-giving produces snowballing cycles of debt that destabilize the entire social structure. This analysis suggests the totally fantastic basis to an Anglo-American liberalism built around ideals of private property and well-delineated utilitarianism.

Mauss further interrogated the supposedly neutral technical foundations of social democratic thought ascendant among his professional peers. He traced its contours to efforts in ancient Roman contract law to rationally calculate debts and responsibilities. From these sources emerged a Western myth of autonomous individuals whose reciprocal ties could be tamed through mathematical tallying. This tactic was not without its benefits. Mauss contended that modern societies had tamed wildly fluctuating patterns of gift-giving by refashioning humanity as homo oeconomicus at the cost of reducing ethics to actuarial calculations. The modern industrial state and its vast apparatus of technological, scientific, and rational adjudication was an offshoot of these fabulous inventions. It was only recently, he wrote, that man had become a "machine, made complicated by calculating machines."[21] Liberal juridical constructs based on self-possessive individualism as well as on the natural and human sciences in their service were complicit in this transformation of humanity into machinery, society into a system of well-regulated inputs and outputs. Mauss categorially dismissed this notion, portraying society as an agonistic affair that no technics could neutralize, hinting that that policymakers ignored these dynamics at the commonweal's peril.

Mauss's study of the gift presented his skepticism toward the liberal individualism present in anglophone political philosophy from John Locke to the present day.[22] The Rockefeller Foundation's initiatives, by contrast, were grounded in a commitment to modernization based on improved technocratic and rationalist social engineering, the division of social problems into tractable data sets, and ultimately the cultivation of liberal individualist subjects who, through their bootstrapping enterprise, would contribute to their community, economy, and nation. Their grant-giving presumed that experts in the United States, empowered by the largesse and reason of their benefactor, could freely identify "sectors" for scientific reform and as such empower exceptional scientific individuals to liberate reason from tradition. Though suited to the privatized, localized, and decentralized networks of American higher education, it proved ill-suited for the rigid, centralized, and techno-statist framework of French higher education. Moreover, couched within these activities were covert political and philosophical assumptions about the constitution of science and the framework for reason itself, which varied according to the American and French traditions. Reforming French universities demanded a more thoroughgoing rearrangement of French and US scientific relations.

The École Libre as Methodological Crucible

Lévi-Strauss's attraction to cybernetics followed his work in the 1940s at the French university-in-exile, the École libre des hautes études, where Russian linguist Roman Jakobson famously initiated Lévi-Strauss into structural methods. Under Jakobson's influence, Lévi-Strauss ceased to study the empirical facts of indigenous kinship and focused instead on the relations among terms that constituted a kinship system proper. A welter of data points transformed into symbolic relations, operating according to a principle of communicative difference suggested by structural linguistics. With the aid of a French mathematician, Lévi-Strauss even found algebraic expressions for his kinship studies. Moreover, the structural analysis embraced by Lévi-Strauss offered a convenient framework for grasping ethnographic facts as diverse as trade, myth, and kinship as relational terms available for computational and communicative study. The result was not merely a leap in anthropological method. It also sketched a possible incorporation of Continental anthropology into the computational analytics of the emerging behavioral sciences in the United States. The ELHE proved the decisive hub in these transformations: the place where the personnel, methods, and technical precision of diverse disciplines fell into line with an emerging style of US technocratic research.

The epistemic transformations set in motion at the ELHE reflected a strategic coup by administrators at the Rockefeller Foundation and the New School for Social Research, who collectively hosted and funded its operations. The political agenda of the ELHE helps explain how its researchers fell under the sway of cybernetic methods. In the first instance, the ELHE provided refuge to francophone scientists driven into exile by German policies in Europe. That humanitarian mission dovetailed with a political agenda to run the ELHE as an experimental laboratory redirecting francophone science in technocratic directions.[23] By the time of the ELHE's founding, the New School was already hosting an eminent community of German-speaking scholars supporting the Allied effort. New School president Alvin Johnson felt that the rise of the Vichy regime in France justified an additional program for francophone intellectuals. Sensitive to the agenda of the Rockefeller Foundation beyond wartime humanitarian activities, he cast the program as an auspicious occasion to initiate younger French scholars into Franco-American collaborations. He proposed what Kittredge characterized as "collaborative research programs related to world problems of economic, political, legal and social character."[24] Success required participants to eventually repatriate and reform governance in their home countries. They promoted this goal by carefully vetting applicants, giving priority to younger candidates likely to have long-term influence at home (a factor, it seems, in Lacan's summary rejection).[25] Fellowships were limited to a few years on the presumption that Allied success would allow scholars to return home, bringing with them scholarly innovations from US science.

Faculty and students consulted for the US State Department, General de Gaulle's government-in-exile, and other anti-Axis agencies. The faculty of law and politics led courses on the illegality of the Vichy regime.[26] Psychoanalyst Raymond de Saussure (the nephew of Ferdinand) trained French women to serve as social workers during the reconstruction of France. Lévi-Strauss drafted proposals for planning the postwar economy in France, while Boris Mirkine-Guetzevitch, an eminent constitutional scholar, convened a commission aimed at rewriting the French constitution according to American legal norms.[27] A summary prepared at the war's end explained how the committee's initial focus on constitutional problems had gradually expanded to ask what some members considered a still more fundamental question: Why is America so efficient? The report explained, "The war has shown the tremendous economic strength of America and its ability to maintain a high standard of living even in wartime. No other nation can offer such an example. If France wants to reach a reasonable level of well-being, she must know why and how the United States achieved such good results."[28] This emphasis on efficiency underscored the success of the Rockefeller Foundation in nudging French scholars away from a concern for social equality. In identifying well-being with a high standard of living, the committee had overtaken core assumptions of Rockefeller reforms, which saw a robust industrial economy, rather than state intervention, as the mechanism for resolving inequity.

Lévi-Strauss's intellectual path until his arrival in New York prepared him well for life at the ELHE. He spent much of the 1920s and the early 1930s as an active socialist, friendly to Marxism, crusading at one point against the evils of colonialism.[29] He put those political interests aside to pursue ethnography. From 1935 to 1939, before commencing his PhD thesis, he undertook ethnographic research and teaching in Brazil on a French-sponsored mission subsidized by the Rockefeller Foundation. These efforts tied in with needs of colonial administration outlined by Mauss, but they also meshed with a program of French scientific diplomacy that cultivated imperial prestige by staffing the Global South with French experts. The collapse of French forces and the establishment of the Vichy regime, however, interrupted Lévi-Strauss's plans for graduate study. Anticipating that it was only a matter of time before he found himself in a concentration camp or worse, in 1941 Lévi-Strauss fled France on a steamer bound for the United States via Martinique.[30]

Lévi-Strauss's fraught flight to the United States captures the treacherous balance of scientific and political interests facing refugee intellectuals. He offered the canonical—and, my research shows, misinformed—account of this conflict in his memoir, *Tristes Tropiques*. On a ship bound for the Americas, he carried with him a trunk filled with "linguistic and technological files, travel-journals, fieldnotes, maps, plans, photographic negatives, thousands of sheets of paper, filing-cards, and rolls of film," which he knew he must keep from US customs agents.[31] "The vocabularies," he later explained, "would certainly strike them as an elaborate system of codes, and the maps, plans, and photographs they would interpret as pieces of military information."[32] After traveling to Puerto Rico by a second boat, Lévi-Strauss was mistaken by border officials for a Vichy emissary. Lévi-Strauss reports that when an FBI official arrived to inspect his person and things, "I ran to the Customs house and threw open my trunk. A solemn moment! He was a well-mannered young man, but when he took up a card at random his face clouded over and he spat out the words: This is in German! It was, in effect, a note drawn from the classic work of von den Steinen, my illustrious and distant forerunner in the Mato Grosso: *Unter den Naturvölkern Zentral-Braziliens*, Berlin, 1894. The long-expected specialist was reassured to hear of this, and before long he lost all interest in my concerns, and I found myself free to enter the U.S.A."[33] So it was that Lévi-Strauss, writing in *Tristes Tropiques*, reconstructed a perilous journey to the United States that was nearly interrupted by his precious cargo of ethnographic data.

The true cause of the interruption related to a system of political information processing governing US immigration in the 1940s. Shortly before his arrival to the United States, an anonymous informant from Poughkeepsie, New York, wrote a postcard to J. Edgar Hoover, identifying Lévi-Strauss as part of a cabal of "Jewish international communists" and thus bringing him to the attention of the FBI.[34] It appears that intervention by the New School sufficed to secure safe passage, but he would continue to be intermittently monitored by the FBI for close to another decade. After he arrived in New York and assumed his post at the ELHE (Centre d'études et d'informations pour les relations avec l'Amérique centrale et l'Amérique du sud) (CCLAR), FBI agents began intercepting his mail and making inquiries in New York. They scrupulously inventoried Lévi-Strauss's prewar undertakings in South America, his work at the ELHE, and his various activities consulting and broadcasting speeches for the US government. Hoover did not like what he saw. In one ominous memo, he noted a recent informant's claim that Lévi-Strauss and surrealist André Breton, both of whom supported wartime propaganda services against Vichy, were "closely connected with a group in Mexico which is very bad, having something on their minds different from what the rest of us have on our minds."[35]

Lévi-Strauss capably reconciled himself with the US security state, and its voracious appetite for technocrats, at the ELHE.[36] He was swiftly appointed director of the ELHE's Center for Central and Latin American Relations (Centre d'études et d'informations pour les relations avec l'Amérique centrale et l'Amérique du sud). The center provided intellectual support for the US Office for Inter-American Affairs directed by Nelson Rockefeller (whose grandfather endowed the main Rockefeller charities). Lévi-Strauss's responsibilities included teaching, convening panels, and writing intelligence reports on Latin America. This activity touched on strategic concerns from political subversives in Brazil to challenges integrating indigenous people into a Latin American civilization.[37] He also consulted with the Office of Strategic Services (oss) and recorded French translations of President Roosevelt's speeches for broadcast into France by Allied propaganda services.[38] These efforts allied political intervention with academic expertise and media consulting. Lévi-Strauss's confidence as political operator hinted at administrative ruthlessness impatient with liberal deliberation, as when he advised an oss agent that summary execution of France's 50,000 alleged Vichy collaborators might be preferable to protracted postwar trials.[39]

Over the latter half of the 1940s and well into the 1950s, through myriad meetings such as those with the oss, Lévi-Strauss applied these finely honed political skills to cultivating exchange between France and the United States. Following the landings in Normandy in 1944, de Gaulle's government summoned Lévi-Strauss to Paris to represent the state in helping French intellectuals visit the United States. War had left the French universities in disrepair and intellectually isolated. By dint of his wartime experience and avid loyalty to de Gaulle, Lévi-Strauss was deemed ideally suited to rearticulate relations and exchanges between the two nations' universities. In 1945, he returned to the United States as cultural attaché to the French embassy, where he continued in a similar capacity, assisting the likes of Jean-Paul Sartre, Simone de Beauvoir, and Albert Camus.[40] As a representative of the ELHE, Lévi-Strauss, along with physicist Pierre Auger, met with Rockefeller Foundation officers to discuss plans for the future of the ELHE. Lévi-Strauss proposed re-establishing the ELHE as a new center in Paris to be called "the French-European American Foundation."[41] Lengthy negotiations and additional support secured from the Ford Foundation resulted in the establishment of the long-sought but never before realized Sixième section of the EPHE, later renamed École des hautes études en sciences sociales (EHESS).

How did Lévi-Strauss succeed at securing a rapprochement among American philanthropies and French institutions exactly where his eminent predecessor Mauss had failed? Commenting on the origins of the school, latter-day faculty member Pierre Bourdieu once complained that the EHESS was an instrument of "social control" deployed by American foundations to counteract Marxist criticism.[42] Speaking more directly to the nature of that control, the French geographer Jean Gottmann, who had close ties to the Rockefeller Foundation, once noted that staff at the Sixième section were viewed as "representatives of the American millions attempting to conquer French thought."[43] This was indeed a strategy of political transformation of the sort that would become a pillar of American "nation building" in decades to come. If it attracted such ire among French intellectuals, this was in part because it so closely resembled France's own policy toward colonial subjects worldwide. In France now, as in Algeria, Senegal, and Indochina before it, the physical control imposed by battalions found a cultural counterpart in the infrastructure of administrators, technocrats, and methods interfacing with the local population. Indeed, if local people of solid character and education could be enlisted, all the better. Where Mauss could facilitate the realization of such infrastructures outside France, Lévi-Strauss could support that realization within. He and a large contingent of colleagues had cultivated relationships, methods, and purposes throughout the war that established the possibility of a new partnership between a robust program of social-scientific technocrats in the United States and the acutely felt need for comparable cadres in French reconstruction. Indeed, Levi-Strauss's elaboration of informatic theories for the social sciences seems to wax in direct proportion to his accumulation of administrative and technical roles, culminating in the early 1960s when he founded the LSA in Paris.

In his embrace of cybernetics and accumulation of technocratic roles, Lévi-Strauss followed a well-trodden path already marked out by his acquaintances Mead and Bateson, who had spearheaded cybernetics with funding from the Josiah Macy Jr. Foundation and the Rockefeller Foundation. Where Mauss elaborated social theories that seemed to sabotage the epistemological and administrative aims of American technocracy and its handmaidens in social science, Lévi-Strauss seized on the language of cybernetics and preached its message at major international conferences. In the hands of Lévi-Strauss, social theory, technocratic administration, and intercultural mediation aligned. Perhaps more remarkably, their alignment took place under the sign of a cybernetic Mauss, whom he construed as a cybernetician *avant la technique*. Much as Mead and Bateson mutually adapted the Culture and Personality school of anthropology to the emerging frameworks of cybernetics and information theory, Lévi-Strauss would come to read Mauss as a protocybernetician. This was an administrative and scientific maneuver, to be sure, but it was also a political maneuver— that is, it took a position on how fundamental conflicts in society were to be defined, endorsing scientistic solutions whose truth resided in the purity of mathematics and machinery rather than the messy contingency of historical reason.

The Stirrings of Cybernetic Structuralism

According to Mauss, there is a virtual dimension to every gift. The physical object and its exchange take for granted a network of past and future exchanges, within whose networks the circulation makes sense. A single local exchange means nothing outside these ties that not only bind, but structure possible futures of production, be they in the realm of knowledge, politics, kinship, or texts. Yet recognition and service to that structure varies tremendously. It is a small testament to midcentury cultures of exchange that Jakobson reciprocated Lévi-Strauss's 1949 gift of a "very stimulating" book on Arthur Rimbaud by sending him a copy of Wiener's *Cybernetics*.[44] Not long after, at the behest of Jakobson, mathematician Warren Weaver of the Rockefeller Foundation dispatched exemplars of *The Mathematical Theory of Communication* to Lévi-Strauss and to French psychoanalyst Lacan.[45] In discourse networks of 1949, French symbolist poetry and thinking machines occupied common portions of a large and amorphous territory of textual analysis. As Shannon and Weaver's book had appeared simultaneously in a French edition in Paris, Lévi-Strauss responded that he had already read it with enthusiasm, prompting him to ask of Jakobson, "Do you think that the apparatuses he [Wiener] describes could calculate a priori all possible phonological structures, permitting the identification of disappeared or poorly known languages?"[46] With this remark, Lévi-Strauss gave tentative expression to a concern that would animate much of his work in the coming two decades: namely that the symbolic sciences of cybernetics, information theory, and game theory could deduce the scope and patterns of structural communication in fields including language, kinship, and commerce.

The first record of Lévi-Strauss's public remarks on using cybernetic means to reconstruct ethnographic data appears in the diaries of Rockefeller Foundation officer Charles Fahs, who wrote in September 1949 that he had traveled to the Conference of Americanists (ethnographers of the indigenous peoples of the Americas) "primarily to hear the paper of Levy-Strauss on the relevance of cybernetics to research in linguistics."[47] Lévi-Strauss opened the talk by disputing Wiener's claim in *Cybernetics* that social science lacked stable, reliable data sets for cybernetic analysis. Lévi-Strauss pointed to written language as a counterexample. He expanded this into his signature tripartite structural and cybernetic rereading of linguistics, economics, and kinship. As Weaver had proposed in *Scientific American* two months earlier, Lévi-Strauss argued that engineering models of communication could be transposed onto all other fields of human activity, including linguistics, economic transactions, and the circulation of women within primitive systems of kinship. According to Lévi-Strauss, these activities composed systems of communication whose terms of exchange (phonemes, goods, and wives) invited analysis by computers.[48] This claim of a logical structure that could be reconstructed, even for "disappeared or poorly known languages," spoke directly to the central challenge facing Americanists. In effect, he shifted the location of culture and the task of the human sciences from physical artifacts processed and witnessed by human minds to abstract models whose reality was attested to by machines.[49]

Lévi-Strauss's proposal to unearth structural relations fit within a larger program of efforts to reconstruct devastated cultures by techno-rational means. In the first half of the twentieth century, ethnologists responded to the evisceration of indigenous life by collecting the artifacts of their disappearing ways of life. By the late 1940s, the order of things once built around ethnographic traces and their exhibition no longer held. National Socialism, the Holocaust, and postwar occupation shook French intellectuals' faith in civilizational hierarchies defined by the historical sequence of material artifacts. Auschwitz, Hiroshima and Nagasaki, and the ongoing destruction of indigenous populations—a trend that decolonization did not reverse but certainly made less available for European observation—heralded the instability of physical indexes in the modern world. Mathematical constructions of reality in computing, atomic physics, operations research, and game theory figured centrally in the unheralded violence of World War II. In the hands of the state, modern science could order or annihilate the human body at scale with relative ease. "Structure" offered a logical category that subtended the physical body. References to structure proliferated in the years to follow: syntactic and grammatical structure, molecular and atomic structure, DNA structure, the structure of scientific revolutions, social structure, structural-functional sociology, the structure of social action, plans and the structure of behavior, and structural architecture. A new order of things was coming into being, disclosed by mathematical deduction.

Awaiting a day when even an anthropologist could share time on an IBM computer, Lévi-Strauss laid groundwork for human sciences equal to the task. The Sixième section availed a faculty agreeable to these aims. Its tasks included theorizing the human sciences in terms suitable for the scientific transformations of the day. Lévi-Strauss's idiosyncratic 1950 introduction to the collected works of Mauss offered his first published foray into communication theory. In passing remarks, he reinterpreted Mauss's *Essai sur le don* as proof that "the ethnological problem is a problem of communication."[50] Recall that Mauss contrasted the exchange of gifts in primitive society to the highly technical and mathematical schemes of modern liberalism. Lévi-Strauss boldly conflated these two positions, arguing that the binding together of primitive societies by means of gift-giving manifested a communicative structure of the sort implied by cybernetics, game theory, and information theory. He dismissed contradicting positions in Mauss's work as prescientific errors.[51] Lévi-Strauss invoked Jakobson's recent work as a resource for tidying up Mauss's oversight. "Social anthropology," he wrote, "can hope to benefit from the immense prospects opened up to linguistics itself, through the application of mathematical reasoning to the study of phenomena of communication" such as cybernetics and information theory.[52] Problems of ethnology and sociology, he explained, were merely "waiting upon the goodwill of mathematicians."[53] Communication theories for recovering noisy signals became models for fighting disorder, noise, and contingency in social-scientific data. This provocative reading transformed Mauss's most contentious political claims into problems in data processing available for resolution with proper computation.

The Administration of Data and the Administration of Persons

In the introduction to Mauss's work, Lévi-Strauss broached other instruments crucial to the rise of structural anthropology. The United Nations Educational, Scientific and Cultural Organization (UNESCO), according to him, could do a tremendous service by carrying out Mauss's proposal to inventory "techniques of the body" around the globe.[54] Such an inventory would, Lévi-Strauss asserted, "be of truly international benefit," capable of drawing on all the world's knowledge in identifying "possibilities of the human body and of the methods of apprenticeship and training" involved in their cultivation.[55] This expressed Lévi-Strauss's blooming interest in anthropology as an aid in a highly technologized postwar order. Ethnographers working for global nongovernmental organizations would guide development while validating subjugated peoples' knowledge. An encyclopedia of techniques of the body promised indigenous technologies for governance at precisely the moment when fascism, the Holocaust, and atomic weaponry had shaken faith in the benevolence of modern and Western methods. In Lévi-Strauss's view, UNESCO offered the vehicle for bringing these new political techniques to the world.

The invocation of UNESCO highlights how Lévi-Strauss envisioned that the administration of data and the administration of populations coincided in the gestating program of structural anthropology. This called for technologies, theories, and organizations, all of which UNESCO might help to develop. It was not enough, however, merely to develop new instruments of social-scientific reason. It was also necessary to reorder that reason through organizations capable of incorporating indigenous forms of reasoning into their outlooks. In this effort, mathematical sciences— particularly those associated with cybernetics, information theory, and game theory—were to play a double role of reforming social science and validating primitive thought. Sciences such as cybernetics modeled how apparently haphazard practices of primitive cultures could be understood as logical systems. From his position in postwar France, where scientific funding remained scarce, UNESCO seemed a promising vehicle for these efforts.

The high modern social science championed by Lévi-Strauss found epistemological, aesthetic, and institutional embodiment in UNESCO's Paris headquarters (figure 4.3). Its spaces, agendas, and art collection (which placed works by modern masters such as Braque and Picasso alongside works by anonymous craftspeople originating from Papua New Guinea, Samoa, and elsewhere) seemed to promise a synthesis of modernism and primitivism. It offered a concrete corollary to Lévi-Strauss's insistence well into the 1960s of complementarity between informatic and indigenous forms. Historian Teresa Tomás Rangil characterizes UNESCO as practicing a "politics of neutrality" that reduced cultural diversity to elemental scientific principles stripped of political specificity.[56] In an effort to offend neither the United States nor Soviet blocs, UNESCO consultants dissolved concrete differences—cultural, ethnic, political—into formal properties of a universal human condition. This analytics comprised a political and

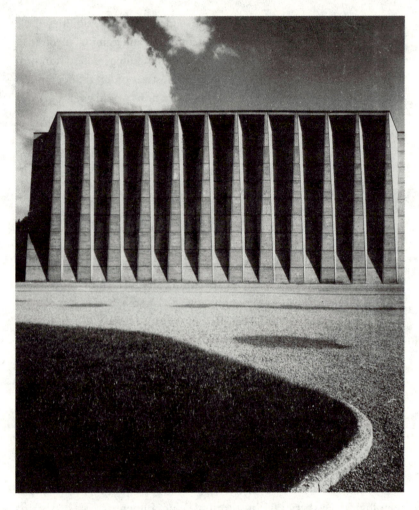

4.3 Side view of UNESCO Conference Building in Paris, photo by Lucien Herve. Source: Barry Bergdoll, "Marcel Breuer and the Invention of Heavy Lightness," *Places Journal*, June 2018.

aesthetic technique, on display in the very architecture of Maison de l'UN-ESCO, which opened in Paris in 1958. A modernist masterpiece designed by Bernard Zehrfuss, Marcel Breuer, and Luigi Nervi, it staged a vision of global unity realized in a stunning array of poured concrete and intersecting grids—as if the challenges of first and third worlds, capitalist West and communist East, might be resolved in mathematically defined lines drawn in a single space.[57]

Lévi-Strauss elicited from cybernetics, information theory, and game theory an epistemology in harmony with UNESCO's depoliticized globalism.

Biographer Emmanuelle Loyer notes, "his American contacts, his connections with foundations, his interest in scientific projects with theoretical and systematic ambitions" ideally matched UNESCO's aims.[58] In 1949 the organization invited him to write a declaration on racism and to supervise a study in rural France on tensions between social groups.[59] He accepted the invitation, telling one Rockefeller Foundation officer that UNESCO's commitment to world peace would "contribute to the development of the Social Sciences Section in strategic ways."[60] In 1953 he was appointed general secretary of its International Social Science Council, thanks partly to his experience navigating the global scientific networks of robber baron philanthropy in which UNESCO hoped to intervene. Writing for UNESCO's *International Social Science Bulletin*, Lévi-Strauss cited research in cybernetics and information theory, structural linguists' emerging collaborations with engineers, and machine translation as exemplary instances of a new qualitative mathematics appropriate to the organization's global agenda. "Specialists in such apparently distinct fields as biology, linguistics, economics, sociology, psychology, communication engineering and mathematics have suddenly found themselves working side by side," with a common language that bridged "capitalistic economics and Marxist economics."[61] Structural anthropology, he believed, would be the great synthesizer of these scientific, political, and cultural differences.

Within Lévi-Strauss's embrace of UNESCO, however, reappeared the old ambivalence of the human scientists in the employ of state technocracy.[62] In the case of Lévi-Strauss, this skepticism was precedented. It found fragmentary expression in his praise, as a student journalist, for Louis-Ferdinand Céline's 1932 novel *Journey to the End of the Night*. He characterized Céline's symmetry of debasement in the developed and developing world as an apt "hatred of mankind, or more precisely, for the society that renders men so detestable."[63] So too his ambivalence toward UNESCO's aims found justification in expeditions to Brazil, whose national motto "ordem e progresso" offered a sour tribute to the French positivism of Comte and Durkheim. Lévi-Strauss wrote in his Brazilian ethnographic diaries that foreign visitors to the tribes of Brazil "cannot but feel anguish and pity at the sight of a people so totally dis-provided for; beaten down into the hostile earth, it would seem, by an implacable cataclysm."[64] These sights cultivated his abiding suspicion that ethnography, and indeed the entire program of European science and education, was coextensive with a will to dominate at home and abroad. Later, writing from wartime New York, he likened political liberalism to an old man, suffering and starved, who persisted in writing noble meditations without any regard for the fact that his "enemies, the opponents of his ideas and his race, are today all powerful, and that he has been left in their hands."[65]

Communication science, with its apparent elision of human agency and appeals to cybernetic entropy and game-theoretical conflicts, suggested to Lévi-Strauss a new account of humankind freed from entreaties to individual freedom and civilizational progress. Many American and even British cyberneticians spoke of cybernetics in triumphalist tones, but Lévi-Strauss found in it dire warnings of communication as a force for exhausting natural resources and cultural diversity. In the years following World War II, Lévi-Strauss increasingly returned to the horrors of colonialism to grasp the proper task of European sciences. He tied these thoughts together in his 1954 memoir *Tristes Tropiques*, where he wrote that "'entropology,' not anthropology, should be the word for the discipline that devotes itself to the study of this process of disintegration" observed by his field's practitioners.[66] Coming at the conclusion of a memoir that recounted not only his time in Brazil but also his flight from Vichy France, *entropology* subsumed the horrors of the Global South and the Global North under a single cybernetic relation. In it, the cruel reduction of indigenous life under Brazilian colonization served as a strange harbinger of the inevitable collapse of France's imperial might in the wake of Vichy collaboration. Within that relation, the anthropologist served as both agent and observer, poison and potential antidote, in communicative decimation.

In a contribution to the 1954 UNESCO volume *University Teaching of the Social Sciences*, published at a time when the organization was championing global literacy initiatives, Lévi-Strauss warned against unchecked optimism over the benefits of accruing knowledge. "The international organizations, and particularly UNESCO, have so far entirely failed to appreciate the loss of personal autonomy that has resulted from the expansion of the indirect forms of communication (books, photographs, press, radio, etc.)."[67] This counterintuitive, even iconoclastic perspective sketched a theme that would become a cornerstone of the racially and colonially charged orality and literacy debates associated with Marshall McLuhan's media theory in the 1960s. Lévi-Strauss supported his analysis by citing Wiener's claims that larger communities embodied diminished networks of information. In multiplying the size and relays of society, technical communication transformed organic social community into a relation of "senders" and "receivers" whose reality was "hidden behind the complex system of 'codes and relays'" that held them together. The analysis, again, relied on entropy: in Wiener's description, society takes the form of an informational system that becomes more disordered as local face-to-face relations give way to an overload of relays.[68]

In Lévi-Strauss's 1952 UNESCO booklet *Race and History*, he elicited from game theory a full-throated assault on the notion that liberty entailed free association, free movement, and free expression.[69] Arguing that diversity and difference were fundamental cultural values from which other cultural values emerged, he argued for the logical necessity of maintaining separation and difference between peoples. He supported this with a game-theoretical analogy that imagined cultures as players in a game of innovation through original combinations of distinctive cultural repertoires. As the game went round and round, with players making innovative creations in collaboration, the two players adapted elements from one another's cultural repertoire, subsuming and finally extinguishing their differences. This brought the game to an end, both players having lost. The analysis recapitulated (without citing) Wiener's thermodynamic account of how distinct environments, brought into communication, gradually exhausted one another through exchange. This bequeathed to international institutions such as UNESCO the task of suppressing "diversities now serving no useful purpose, the abortive remnants of forms of collaboration whose putrefying vestiges represent a constant risk of infection to the body of international society. They will have to cut them out, resorting to amputation where necessary, and foster the development of other forms of adaptation."[70] In a sense, this recapitulated Céline's suggestion that global modernity debased colonizer and colonized alike. To keep the game going, to ensure cultural vibrancy, intermingling had, somehow, to be checked. Some manner of prohibitions, a restriction on communication, seemed necessary to a robust postwar world.

These and other writings of the 1950s displayed Lévi-Strauss's idiosyncratic fidelity to the technocratic ambitions the Rockefeller Foundation harbored for veterans of the ELHE. His methods offered to transpose many of the most contentious political issues in postwar France (gender, economic and technological modernization, relations in and with the third world) into highly technical frameworks for expert analysis. They substituted political debate for a cool analysis powered by mathematics and, perhaps, instruments like the digital computer. Against Mauss's warnings against man as a machine—made all the more complex by calculating machines—Lévi-Strauss discovered information-theoretical calculations run amok. Lévi-Strauss even found in UNESCO a vehicle for global social science modeled on the cybernetic, informational, and game-theoretical methods championed by Rockefeller officers. Yet his implementation of these lessons diverged from the vision of the Rockefeller officers who found in him such a congenial partner. Firsthand encounters with the human cost of colonialism and the imperial wane of post-Vichy France revealed a world collapsing under the weight of modernity, in which communication science was both an instrument and an analytic of this destruction.

Imperial Data

Throughout the 1950s, Lévi-Strauss sought to establish a physical infrastructure equal to the tasks of his emerging structural anthropology. Paris disposed of the ethnographically minded Musée de l'homme, which held artifacts from the interwar expedition to Brazil. Its positivist and encyclopedic inventory of cultures, however, was ill-suited to the project he had in mind. In an acid critique, French sociologist Roger Caillois suggested ethnography was not about playing games, like roulette, but about putting together pieces of a puzzle that was the total expanse of humankind. Lévi-Strauss's 1955 retort to Caillois captured his discontent with traditional ethnographic museums. "There exists a major difference between roulette and puzzles: In the former, play is never complete, while . . . in the latter every piece has its final place."[71] The human sciences could never produce a total archive of cultures, for they were defined by virtual relations—structures—rather than physical artifacts. Lévi-Strauss's structural frame of mind suggested the need for something like a database for computing and modeling dynamics. The museum might be a resource in this analysis, but its tendency to isolate exemplary artifacts, or worse yet, install them in a hierarchy of human progress, threatened to obscure the cultural structures essential to their signification. The order of the day, therefore, was interdisciplinary cooperation: ethnologists working alongside mathematicians and engineers, with mathematical techniques poised to uncover the symbolic series governing cultural transformation. This was less an archive for physical artifacts than an information processing center.

Another article published in 1955 outlined Lévi-Strauss's practical effort, underwritten by the Ford Foundation, to devise cultural analytics that went beyond the ethnographic museum.[72] On note cards, he had broken myths into constituent sentences, which he then arranged into regionally specific bundles of relations (life and death, sex, marital status). This analysis, he claimed, permitted consideration of "all the known variants of a myth as a series forming a kind of permutation group," in which narrative was transformed into logical operators.[73] He acknowledged the crudeness of his computational means, regretting that "with the limited means at the disposal of French anthropological research no further advance can be made."[74] Selecting, encoding, and cleaning data often proved difficult. Mythological literature was "bulky," and he warned that "breaking it down into its constituent units, requires team work and technical help."[75] A single iteration of a myth required hundreds of cards and a makeshift analog computer—little more than a bulletin board with pegs for shifting units around—before its computational patterns could be sketched. "Furthermore, as soon as the frame of reference becomes multi-dimensional (which occurs at an early stage, as has been shown above) the board system has to be replaced by perforated cards, which in turn require I.B.M. equipment, etc."[76] He advised, "It is much desired that some American group, better equipped than we are here in Paris, will be induced by this paper to start a project of its own in structural mythology."[77]

Lévi-Strauss's efforts to establish a laboratory for communication research seemed perpetually flustered. Appeals to Le materiel téléphonique, the French homologue of Bell Labs, went nowhere.[78] The Rockefeller Foundation, the Ford Foundation, and UNESCO showed little sympathy for his requests for laboratory funding. French information theorist M. P. Schützenberger, who would later join MIT's Research Laboratory of Electronics, did some outreach on his behalf as well. In November 1951, Schützenberger wrote to Wiener, informing him of Lévi-Strauss's efforts "to set up a center of research on the applications of the theory of communication" to fields such as musicology and mythology. Commending Lévi-Strauss's intuitive grasp of what could be done with cybernetics, Schützenberger added, "he put the thing more or less on my shoulders for he had heard that you trusted me."[79] With Jakobson's help, Lévi-Strauss secured $2,000 from MIT's Center for International Study (CENIS), a center of cybernetic research covertly linked to the CIA, in order to organize an interdisciplinary seminar on cybernetics in Paris.[80] CENIS director Max Millikan, formerly the director of the CIA's Office of National Estimates, must have found in the project an impressive opportunity to cultivate the center's network of international researchers tilting toward American science.

Levi-Strauss's 1960 ascension to a chair at the Collège de France, and his concomitant establishment of the LSA, presented him with the long-sought opportunity to establish a research laboratory. Instruments, students, and researchers—the latter including members of staff and distinguished visitors—filled its ample space. His seminar promptly moved to the new space, where researchers, Lévi-Strauss proudly reported, investigated "the use of descriptive codes, punched cards, and computers in our disciplines."[81] One participant, Jean-Claude Gardin, would later found a national center applying computers to the analysis of ethnographic findings encoded on SELECTO punched cards. An alliance with EPHE professor Jacques Bertin's Laboratoire de cartographie furnished Lévi-Strauss with continuing means for realizing his analysis according to the most current techniques in graphical information processing.[82] The term *social anthropology*, as opposed to *ethnology* or *ethnography*, denoted its founder's plans to not merely inventory facts of ethnic groups but to pursue a systemic scientific analysis of the social dimensions of human systems. The crowning jewel of this new installation, however, was the ample infrastructure of the HRAF (figure 4.2), which, as of 1961, comprised 2 million index cards filling 380 cabinets, weighing 7.5 tons in total and occupying 18 cubic meters. Lévi-Strauss took pride in having secured Europe's only HRAF installation. He proudly availed qualified colleagues with access to the files, including Lacan, philosophers Raymond Aron and Simone de Beauvoir, and sociologists Gabriel Le Bras and Pierre Bourdieu.[83]

In founding the LSA, Lévi-Strauss realized a new form of social knowledge precisely attuned to the political infrastructures of 1960s global science. The laboratory's knowledge of the world, the forms of data and representation it lent itself to, reflected a distinct and recent history of empire. Founded at Yale University in 1935 as the Cross-Cultural Survey, its original purpose was to provide a systemic overview of comparative research in anthropology. During World War II it was repurposed as a tool of US military intelligence. Government clients included the Office of Inter-American Affairs directed by Nelson Rockefeller and supported by Lévi-Strauss in his wartime directorship of the ELHE's Center for Latin American Studies.[84] Between 1943 and 1944, the US Navy relied on the survey to compose at least twenty strategic studies of territories it planned to occupy across the central Pacific.[85] One such publication, the *Military Handbook on Kurile Islands*, explained that facts were "selected for their pertinence to the problems of administrative planning and action."[86] It covered such issues as racial characteristics (the indigenous Ainu population was labeled "docile, submissive, and spiritless"), militancy ("In the distant past the Ainu were warlike; today they are completely subdued"), and sexual customs (the Ainu "enjoyed complete freedom in sexual relations before marriage").[87] Other volumes followed the same format. Wartime exigency had transformed an unwieldy mass of ethnographic observations into an agile information system accommodated to standard classifications. The initial searchability of that data as so many military handbooks, equipped with standardized titles and categories across diverse locales and peoples, coincided with the reach of a Cold War empire in the process of being born.

The HRAF was the enduring expression of political information processing. It was part of the same technocratic quickening that systematized anthropology during World War II with the cybernetic methods embraced by practitioners including Lévi-Strauss, Mead, Bateson, and Clyde Kluckhohn. This synthesis gave rise to a bold cultural analytic whereby vast regimes of human data were disassembled into informational units for cross-cultural analysis. Knowledge scattered across thousands of peoples, millions of texts, and hundreds of millions of bodies would now assume the form of orderly comparative data. The HRAF applied strategic urgency to empowering ambivalent human scientists in their struggles to understand remote and disappearing peoples. This was not Caillois's puzzle of consolidating a full account of humankind of which Western nations were at once pinnacle and steward; it was instead von Neumann's agonistic theory of games, modeling and planning for strategic combinations. In it, a superpower's effort to master a global terrain coincided with human scientists' dream of capturing disappearing cultures. Allied, they gave rise to cultural sciences preoccupied with simulations—strange simulations whose implications in the real historical collapse of indigenous cultures trained their analysts' minds not only on culture and data but also on new problems like simulacrum, pastiche, and hyperrealistic knowledge less anchored in material referents. They were part of a global apparatus of knowledge that, paradoxically, unmoored cultures from local and embodied reality. In short, in the HRAF and the anthropological laboratory, as in cybernetics and the LSA, a nascent postmodern condition took shape under the systematizing power of political information processing on a global scale.

When Marxist historian Maxime Rodinson suggested that structuralism was little more than US imperialism dressed up as French social theory, Lévi-Strauss's response suggests how he might have answered political critiques of the LSA. "Mr. Rodinson's attitude," Lévi-Strauss wrote, "certainly corresponds to that of an orthodoxy which has asserted itself boisterously with respect to linguistics, physics, biology, and cybernetics."[88] Lévi-Strauss asked in reply, "Should we not distinguish scientific findings, strictly speaking, from the political and ideological uses to which they are put, all too frequently, in the United States and elsewhere?"[89] Confronted with these critiques, Lévi-Strauss demurred, preferring to turn the inquiry to finer details, such as exogamy in kinship or competing accounts of Zuni Indians' emergence myths, to see if they withstood scrutiny. These two closely related methods of defusing political debate—the distinction of findings from their context and the argument for steeping debate in dense and technical disciplinary literature—reflect a single technocratic logic. In it, debate shifts from concrete embodied facts to the capacity to navigate, synthesize, and rearrange relations among data. Its analytical instruments include an accompanying apparatus of seminars, journals, scientific theories, and academic networks empowered to issue authoritative analyses. Against the political passion of the café, structuralism offered the cool analysis of systems, organized around the collective manipulation of symbolic elements, informed by the sciences of communication. This, too, was a political passion.

Five

Learning to Code

Cybernetics and French Theory

[Philosophy's interest in writing today] is due to a certain total transformation . . . that also can be ascertained in other determined fields (mathematical and logical formalization, linguistics, ethnology, psychoanalysis, political economy, biology, informatics, programming, etc.). **Jacques Derrida, "Implications," 1967**

In the years after liberation from the Vichy Regime, French citizens' confidence in their individual and collective agencies seemed shaken. It was not just a matter of coming to grips with occupation and its attendant genocide. Knowledge of their fellow citizens' willing collaboration in the deportation of tens of thousands of Jewish people also figured in the sense of disquiet. In addition, if National Socialism was a certain kind of modernity and bureaucracy run amok, the 1950s were a peculiar relief. Sweeping programs of modernization, administered by tens of thousands of experts and consultants, thrust the largely rural Gallic nation into industrial modernity. A vast network of instruments, techniques, and consultants, mediated by consumer goods and advertising, seemed to adjudicate social difference. Literature and the arts picked up on these unsettling dynamics early on. Nathalie Sarraute conjured their spirit in her 1948 novel *Portrait of a Man Unknown*, an account of dreary depthless life ordered by cheap consumable goods and petty sums of money.[1] In 1957 agricultural-engineer-turned-novelist Alain Robbe-Grillet wrote of individuality and interiority being eclipsed by a new era of "administrative numbers."[2] Swiss artist Jean Tinguely's mechanical "metamatic" sculptures, displayed in Paris, and French director Jacques Tati's films of traditional life disordered by modernist cubes likewise staged human agency as a kind of residue gumming up the automated mechanisms of postwar life (figures 5.1 and 5.2).

In French universities, the question of human agency and technical mediation took the form of a preoccupation with the human sciences or, as structural psychoanalyst Jacques Lacan put it, the "so-called human sciences—which, already, aren't all that human."[3] In a French republic abruptly relegated from great empire to secondary (perhaps tertiary) world power, the systems analysis sweeping social and cultural theory embodied the contradictory prospects of growing technocracy. Structuralism, the theoretical fad storming first linguistics, then anthropology, psychoanalysis, semiology, philosophy, and literary criticism, emerged as a forum for exploring and contesting the new technocracy—embracing its promises as well as condemning its pettiness. Critics considered the structuralist enterprise an agent of US occupation, the exchange of one foreign occupying force for another. As "liberation" delivered armies of Marshall Plan economic consultants to France, French structuralists' talk of codes, systems, and communication seemed to poach, or maybe parody, the jargon of visiting US experts.

This chapter argues that French structuralist theorists' part in the postwar occupation of France was more that of squatters than of soldiers. Structuralists' initial enthusiasm for *culture as communication* gave way, by the early 1960s, to a more sinister preoccupation with *culture as code*. Literature, the press, science, politics, asylums, and the factory became instances not merely of communication and exchange, but more ominously of encoding and decoding. The analysis of these dynamics as part of the fashionably ascendant human sciences reflected a broader refashioning of the human condition as yet one more operational concept for optimization. As I noted in this book's introduction, in 1967 US sociologist Daniel Bell wrote of a new intellectual era characterized by the "the new centrality of theoretical knowledge, the primacy of theory over empiricism, and the codification of knowledge into abstract systems of symbols that can be translated into many different and varied circumstances."[4] Structurally minded thinkers, among them anthropologist Claude Lévi-Strauss; literary critics Roland Barthes and Julia Kristeva; psychoanalysts Lacan, Luce Irigaray, and Félix Guattari; and philosophers Jacques Derrida, Michel Serres, and Michel Foucault, applied the emerging intellectual tools to paradoxical ends. They wielded the new emphasis on theory and codes against the dogmas of humanism, and used it to valorize the kinds of analytical and textual operations long practiced in fields like criticism, psychoanalysis, and ethnography. As for the imperial thrust of these emerging knowledge practices, and their

5.1 and 5.2   Photograph of Jean Tinguely on Champs-Élysées in 1959 by
Robert Doisneau and a frame enlargement from Jacques Tati's
film *Playtime* (1967). Tinguely's Paris sculpture performances
of *Meta-matic no. 17* and Tati's film *Playtime* were among a much
larger collection of literary, philosophical, and anthropological
works recasting human agency in the variously whimsical or
threatening light of automatic machinery.

association with US hegemony, that too they put to a new purpose. In a country that identified modernization with imperial expansion, they flirted with visions of France as a new space of techno-rationalization, the human sciences practicing the kind of rational inventories nineteenth-century imperial ethnographers carried out in the colonies.

For most present-day Anglophone readers, recovering how technocracy and colonial anxiety shaped structuralism and poststructuralism demands bracketing much that we think we know about "French theory."[5] French theory circulated in Anglophone universities as a more or less coherent patchwork tailored to the disciplinary debates of elite colleges and universities in the United States during the 1970s and 1980s.[6] That reception privileged an account of postwar French thought as the "hermeneutics of suspicion," a rereading of Karl Marx, Friedrich Nietzsche, and Martin Heidegger through the lens of Saussurean linguistics.[7] There is much to commend that reading. Nonetheless, it also served as a therapeutic for humanities professors rendered less secure by the intellectual era described by Bell. In effect, "French Theory" took the technocratic language of codification—a distinctly postwar phenomenon—and repackaged it as a recovery of the grand traditions of Continental thought. This reading does an injustice to French theorists' fabulously perverse confrontations with modernization, technocracy, and decolonization. Consider the remarks made in 1972 by Belgian-American literary critic Paul de Man, who identified Barthes as part of an intellectual movement giving Hegel, Heidegger, Freud, Marx, and Nietzsche a French inflection.[8] This reading gave credibility, at universities like Cornell and Yale, to the wildly idiosyncratic Barthes, who made a living amalgamating bits of ethnography and linguistics in his classes at postwar institutes for training graduate students in social science. Yet it also sidesteps questions of technocracy, modernization, and communication theory that figured centrally in the development of Cold War literary semiology. This chapter recovers that engagement by reassembling the apparatus of universities, research centers, theories, and problems that mobilized structural thought in France during the 1950s and 1960s.

Research Center and Research Seminar as Experimental Space

Much discussion of cybernetics in the mid-twentieth century, as now, concerned its promise to usher forth a new era of human-machine communication. Yet its early reception in France, as in the United States, displayed an overriding concern for social science and political management. No mere advance in instrumentation, the new information sciences suggested a revolution in social control. A review of mathematician Norbert Wiener's *Cybernetics* in *Le Monde* in 1948 by Dominican priest Dominique Dubarle hailed the new communication science as a prophetic study of how computing could shape world governance. Referencing Hobbes's *Leviathan*, Dubarle positioned *Cybernetics* as a contribution to natural and political science equally. According to his reading, "The human processes which constitute the object of government may be assimilated to games in the sense in which John von Neumann has studied them mathematically."[9] Perhaps, he added, "it would not be a bad idea for the teams at present creating cybernetics to add to their cadre of technicians, who have come from all horizons of science, some serious anthropologists, and perhaps a philosopher who has some curiosity as to world matters."[10] Dubarle intuited the backdrop of social-scientific reform guiding Wiener's claims and anticipated correctly that this would be the major context of its development. The recognition reappeared widely in French commentaries, where invocations of cybernetics consistently emphasized the problem of social engineering.

Dubarle's recognition of cybernetics as a potential human science aligned with a larger managerial reconstruction of knowledge in France after World War II. In industry as in academia, the research group and the research center emerged as elements of a new analytical machinery. This mode of inquiry facilitated research languages adapted to group work and the analysis of systems. Sociologist Pierre Bourdieu, a firsthand participant in these theoretical and logistical changes, argued that changes in academia mirrored those of industry. He cited the "increased complexity of technology" as a major factor in this change. The costs of research and equipment compelled individuals to abandon their status as "unattached cultural producers or small independent inventors for that of salaried cultural producers integrated into research teams endowed with expensive equipment and involved in long-term projects."[11] The natural sciences acted as both model and test bed for this transformation, with their methods of theorization and organizing labor modeling innovation in the humanities and social sciences. "This process of relative dispossession," Bourdieu wrote, "which first took place in the domain of the exact sciences . . . now affects the domain of the human sciences."[12] Industry owed much of this shift to the Marshall Plan, which organized 500 industrial training missions for 47,000 French missionaries between 1948 and 1958.[13] In academia, new research centers fostered by US robber baron philanthropies likewise provided a key impetus for intellectual reform.

The Sixième section of the École pratique des hautes études (EPHE) stood at the forefront of French universities implementing the new intellectual methods (figure 5.3). A new institution for higher learning with a particular commitment to the tools and methods of social-scientific investigation, the Sixième section owed its emergence to postwar efforts to modernize France, aided by generous funding from American philanthropies. Its rise, and the corresponding growth of opportunities and prestige for social science, belonged to the "radiance" of an increasingly technocratic France; it was tethered to a nascent nuclear industry and to techniques of "management" and "human engineering" imported from across the Atlantic.[14] It modeled aspects of its study program on the social sciences in the United States and relied on funding from technocratically minded American foundations, which was delivered partly thanks to Lévi-Strauss's successful mediation. Quite unusually for French higher education, the university welcomed enrollment by students without diplomas, and longtime director Fernand Braudel conceived of research centers as a central forum for teaching.[15] This format explicitly rejected traditional academic hierarchies and forged an experimental research culture that stretched from classroom to laboratory.

New research centers at the Sixième section included Jacques Bertin's Cartography Laboratory, founded in 1954 (whose computer-aided data visualization furnished diagrams for a number of Lévi-Strauss's publications); the Group for Social and Statistical Mathematics, founded in 1955 by mathematician Georges-Théodule Guilbaud (who had recently written a book on cybernetics and was a collaborator with Lacan and Lévi-Strauss); the Center for the Study of Mass Communications (Centre d'études de communication de masse, or CECMAS) and the Center for Interdisciplinary Studies, both founded in 1960 by Georges Friedmann (with contributions from Barthes as well as from Edgar Morin and Violette Morin); and the Center for European Sociology, founded in 1960 by Raymond Aron with Pierre Bourdieu as an assistant.[16] In 1962, funding from the Ford Foundation gathered the various research centers of the Sixième section under a single roof: the House for the Sciences of Man on boulevard Raspail.[17] These developments set the trend for universities nationwide. Between 1955 and 1965 the number of research centers for social science in France jumped, by one count, from 20 to more than 300.[18] The resistance of some French universities, such as the Sorbonne, to establishing

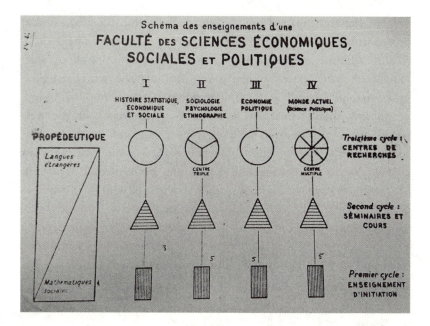

5.3 The Sixth Section of the École pratique des hautes études integrated research centers into the study program (*centres de recherches*) and visualized mathematics and foreign languages as figuring equally in basic training (*propédetique*). From a 1958 proposal by Annales school historian Ferdinand Braudel in the EHESS archives, Louis Velay papers, file "Projet de création d'une faculté des sciences humaines et sociales (juin 1958)." Source: Lucile Dumont, "From Sociology to Literary Theory: The Disciplinary Affiliations of Literature in Section VI of the École Pratique des Hautes Études (EPHE), 1956–1975," trans. Jean-Yves Bart, *Symbolic Goods* 3 (2018): 2.

experimental research centers on this model played into the success of structuralist-influenced methodologies in the human sciences in France.[19] Though well positioned by some measures to link traditional disciplines to new methods, the Sorbonne ceded its opportunity to the newer, nontraditional academic spheres, lending credibility to the new class of theoretically minded scholars' claims to radical innovation.

The new research centers promoted structural and social-scientific methodologies that had had fewer opportunities to develop amidst the sclerotic academicism of more established and elite institutions of French higher education. As the young, growing upstarts of postwar intellectual life, this also made these centers the anchor of rapidly proliferating personnel in postwar French universities. In some sense, they provided a program for the modernization and expansion of knowledge production in postwar

France.[20] Between 1959 and 1969, national professorships doubled, assistant professorships (*maîtres de conferences*) quadrupled, and assisting personnel (*maîtres assistants* and *assistants*) increased by an astonishing elevenfold.[21] Notably, the new patterns of hiring were bottom heavy, with prestigious professorships dwarfed by the proliferation of positions for junior scholars whose methods, reputations, and relations to their new colleagues were not well defined. Moreover, the high proportion of new positions in the social sciences, whose underdevelopment in France was discussed in the last chapter, gave new hires a measure of autonomy from entrenched patterns of academic patronage. Nascent structural concepts, with their emphasis on interdisciplinarity, technical specialization, expertise, and social utility, filled this logistical and even managerial void. Structuralism seemed a forward-looking empirical method, its use of data promising scientific precision and far-reaching advances. This promise attracted younger scholars eager to chart bold yet rigorous programs in emerging research areas.

Research Seminar as Experimental Space

Lévi-Strauss helped define the ensemble of methods, discourses, ideals, and even spaces of collaboration of research centers of postwar France. Exemplary was his research seminar the Utilization of Mathematics in the Social Sciences, taught from 1953 to 1954. It propagated new ideals of interdisciplinary collaboration, cybernetics-themed human sciences, new spaces for intellectual debate, and innovative networks of patronage. Hosted by UNESCO-Paris, where Lévi-Strauss held a position leading research on the social sciences, he funded its activities with subventions from the Ford Foundation and MIT's Center of International Studies (CENIS), both of which had ties to CIA funding.[22] The seminar amalgamated these networks in durable collaboration: CENIS funding came from intercession by Russian structural linguist Roman Jakobson, then working with engineers at MIT. Jakobson secured the funding on a promise that Lévi-Strauss would levy the results toward an edited book on communication and the social sciences for publication in an MIT book series coedited by Jakobson. The seminar crystallized this web of international social relations and alliances cultivated through the agenda-setting that sprang from months of upbeat dialogue around how to fund interdisciplinary collaboration.

Participants in the UNESCO seminar included psychologist Jean Piaget, physicist Pierre Auger, mathematicians Jacques Riguet and Guilbaud, information theorist Benoît Mandelbrot, M. P. Schützenberger (a figure in information theory), psychoanalyst Lacan, linguist Émile Benveniste, cybernetician Ross Ashby, and biologist Remy Chauvin. They presented on topics including "kinship and group exchange," "structure of public opinion," "psychoanalysis considered as a process of communication," "the study of myths as a special form of communications," and "statistics in linguistics."[23] The group consolidated discussions around cybernetics and information theory underway since at least 1951, when Lacan, Benveniste, Guilbaud, and Lévi-Strauss had been privately meeting to develop new mathematical approaches to the social sciences. In a May 1953 letter to Jakobson, Lévi-Strauss described the seminar as striving "to delimit a few problems and find a common language," albeit in a roundabout manner.[24] The seminar also set in motion a long-standing collaboration between Lacan and Riguet, reflected in the former's enduring efforts to spearhead mathematical approaches to structural psychoanalysis.[25] Two other attendees, French engineer François Le Lionnais (director of science education for UNESCO) and mathematician Claude Berge, subsequently cofounded the celebrated avant-garde collective OULIPO (Ouvroir de littérature potentielle, i.e., workshop of potential literature) best known for its algorithmic approach to writing.[26]

The meetings consolidated the growing status of "the seminar" as a site for the clinical analysis of the human condition, remote from the café and man of letters that previously exemplified intellectual authority in France. The postwar seminar was, implicitly, a *research seminar*, a place for the ongoing reconstruction of knowledge and method. Distinct from philosophical or humanistic lectures, it emphasized airing research in progress. Guest lectures and the ongoing synthesis of findings from across fields figured centrally in discussions. Unlike that convivial abode of existentialist thought, the café, the seminar was the domain of the expert. It did not fight for universal values but rather cultivated specialist rigor. Since his appointment to the EPHE in 1950, Lévi-Strauss had held a weekly seminar, often visited by guests back from fieldwork as well as by experts from other fields, including the likes of Lacan and Auger, and eminent scholars from abroad.[27] Where political commitment inspired the café crowd, seminars prized an impartial scientism. Asked once if his political commitments guided his teaching, Lévi-Strauss replied, "Not at all! I am perfectly neutral in my teaching. For me, it is a matter of two compartmentalized boxes. I'm not trying to convert my students. I present the program and only the program."[28] Eschewing political commitments, he said his authority rested "on my work, on my scruples of rigor and precision."[29] Foucault labeled this new kind of postwar thinker a "specific intellectual" whose political responsibility was akin to that of the "nuclear scientist, computer expert, [and] pharmacologist."[30] Their public interventions rested on authority cultivated in laboratories with instruments, equations, and data.

Seminar expertise emerged in concert with a political technocracy that thrived under Charles de Gaulle's Fifth Republic. De Gaulle championed government by experts who were often (like Lévi-Strauss) former communists who had rejected political militancy for the supposedly nonpartisan manipulation of the levers of the state apparatus.[31] These experts wielded interdisciplinarity, mathematical analysis, data, and socioeconomic theory as tools of governance. The UNESCO seminar, with its discursive apparatus of expert knowledge from cybernetics, mathematics, and social sciences, exemplified this emerging analytic. Seminars in a similar spirit proliferated. Consider a 1963–1964 seminar convened by Barthes at the Sixième section, Sociology of Signs, Symbols, and Representations: Systems of Objects (clothing, food, housing).[32] Sessions treated fashion as a rule-bound communication system revealed by theoretical analysis. Participants and contributors hailed from fields including sociology, linguistics, psychoanalysis, and philosophy, with semiology offering a quasi-functionalist framework for their integration.[33] While Barthes's contrarian attitude toward functionalism will be touched on below, his seminars fit within the period's emphasis on experts, systems, and research. The cultivation of a shared intellectual currency in accord with that mode—for example, semiology with its emphasis on codes, systems, communication, economy, and even informatic patterning of signs—relied on the seminar.

Lacan's Laboratory

No seminar exemplified the parallel ascent and critique of technocratic human sciences more than Lacan's. He saw in structural analysis, particularly its penchant for communication theories, an analytic so trenchant, so acidic, it tended toward the evisceration of the human objects it apprehended. His intellectual trajectory primed him for theories that replaced minds with technical relays. His thought was shaped by a generation of thinkers after World War I, among them psychoanalyst Sigmund Freud, for whom trauma and mechanical force shattered old notions of corporeal and psychic integrity. Modernity and technology revealed or produced a body in pieces. In his writings on the mirror stage, first sketched in 1936, Lacan gave voice to these outlooks, suggesting a notion of the body and mind as so irremediably mangled that only strange supplements—mirrors, images, projections—could lend an illusion of integrity.[34] From this perspective emerged a disdain for humanist psychologies that appealed to a supposedly elevated and coherent self. He spoke deprecatingly of a psychology that "discovered ways to outlive itself," well past the waning of credibility of humanism, "by providing services to the technocracy—sliding, as it were . . . like a toboggan from the Pantheon to the Prefecture of Police."[35] (The colorful phrasing alludes to the fact that the main Paris police station sits at the bottom of the hill on which stand two humanist landmarks, the Sorbonne University and the Pantheon entombing Voltaire and Rousseau).

Perhaps Lacan's ambivalence toward technical expertise figured in his summary rejection in 1942 from the same Rockefeller fellowships that had permitted Jakobson and Lévi-Strauss to join the École libre des hautes études in New York and begin assimilating wartime US technocratic thought.[36] Thanks to the postwar mediation of Jakobson and Lévi-Strauss (Jakobson ensured that mathematician Warren Weaver of the Rockefeller Foundation personally sent Lacan a copy of *The Mathematical Theory of Communication*, which Weaver coauthored with information theorist Claude Shannon), Lacan caught up, using his seminar as the forum for reshuffling informatic ideas in the 1950s and 1960s. Here was a kind of technocratic thinking he could get his hands around: a vision of language, culture, madness, and literature as impersonally structured series, cohering here and there in a single mind, but their source powers discerned in equations, diagrams, and computers. The site for propagating his peculiar readings of cybernetics was not a literary seminar nor a philosophy seminar, but a place much more apposite to cybernetics, at least in its US incarnation: a psychiatric hospital. Here was a setting where experts saw that the mind was not a well-rounded soul descended from above, put into human form, but rather a messy, disordered thing, held together by ridiculous masks. In the hospital, Lacan and his colleagues considered how to put together shattered masks, but also theorized the forces sustaining the everyday illusion of psychic integrity. For years Lacan convened weekly sessions in the women's ward at Sainte-Anne Psychiatric Hospital on Wednesdays at 12:15 p.m. so that doctors finishing their rounds might attend. This time and place of meeting aligned the sessions with the medical sciences; it recalls the fact that Lacan and his attendees viewed psychoanalysis as the equal of medicine and biology, a product of the space and methods of the modern research hospital. In 1964, when Lacan shifted his seminars under the umbrella of the Sixième section at Braudel's invitation, the entanglement of scientific and mathematical analytics formed part of their appeal.

Lacan used his weekly seminars as a laboratory for communication science. Freud had shown that for the psychoanalyst, the first and most important object of communication research was the voice—hence, the "talking cure." Its experimental investigation required, in the first instance, little more than a speaker and careful note-taking. Thus was founded the first cybernetic laboratory in France, not in a school of engineering, nor in an industrial laboratory, but in a weekly seminar on Freudian analysis. Biographer Elizabeth Roudinesco called Lacan's seminar "a kind of research laboratory," albeit one populated by "philosophers, psychoanalysts, and writers."[37] Lacan found in psychoanalytic treatment a form of symbolization with parallels in cybernetic and informational theories. He treated the seminar as a discursive technology for theatricalizing these symbolic processes. Through incessant wordplay, diagrams, mathematics, and a dense thicket of references, Lacan turned lectures into a theatrical showcase of symbolization. The priority that Lacan's seminar transcripts often held over his articles reflected the authority of the seminar as a site of knowledge. Like the transcripts of the Macy conferences, they reflected a waxing sense of speech in small groups as the cutting edge of theoretical innovation.

Lacan's infamously abstruse exposition embodied his contradictory relationship to technocratic reason. On one hand, he could evince slavish adherence to an exclusionary vocabulary characteristic of technocratic expertise. Like many high modern theorists in the "age of system," Lacan seemed to relish recasting everyday life theoretical jargon drawn from the technical sciences.[38] Where Freud had often presented psychoanalysis in concrete and widely accessible language, Lacan presented his method in terms favoring the initiated specialist. He shared these linguistic habits with the postwar US behavioral sciences, whose embrace of theory likewise shrouded political matters in an often abstract, technical jargon. Even among fellow structuralists, however, Lacan's discursive style elicited antipathy. In otherwise laudatory remarks, structuralist Marxist Louis Althusser could not resist making a jab: "If you go to Lacan's seminar, you'll see a whole series of people praying before a discourse that is unintelligible to them unless they've frequented the master for a long time."[39] His friend Lévi-Strauss once remarked: "I always had the impression that, to [Lacan's] fervent admirers, 'understand' meant something different from what it meant to me. I'd have had to read everything [he wrote] five or six times. Merleau-Ponty and I used to talk about it and concluded that we didn't have the time."[40] Even Lacan himself seemed uneasy with effects wrought by his discourse, once telling a rapt audience "It's up to you to be Lacanians if you wish. For my part, I'm a Freudian."[41] As with Wiener and Bateson, theorization seemed not merely to produce students but also disciples.

On the other hand, Lacan wielded jargon to distinct epistemic ends, namely the rejection of clear and transparent communication as a positive value. As Foucault put it, Lacan believed that "the obscurity of his writings should correspond to the complexity of the subject, and that the work of understanding the subject should be seen as work to be carried out on oneself."[42] Against one model of cybernetics rooted in frictionless and error-free communication, Lacan insisted on another theory of communication premised on the structural necessity of distortion and ambiguity. The play of meaning in an uncertain phrase corresponded to a specific notion of the linguistic play as central to the concerns of psychoanalysis. That the necessity of reading "five or six times" elicited resistance was precisely his point; structural psychoanalysis worked through this resistance. While aping a certain kind of rationalism, his analysis unraveled the claim to clear and articulate knowledge claimed by technical experts. The result was not really "communication theory" in the field of psychoanalysis but something more like "cacophony theory," conceived as the counterpart to US behavioral sciences increasingly enamored of the idea that noise and errancy might be engineered out of the human condition.

The unabridged text of Lacan's celebrated 1954–55 seminar on Edgar Allan Poe's story "The Purloined Letter" magisterially showcases his skill at reading informatics toward new insights.[43] Overlapping with his participation in Lévi-Strauss's seminar in UNESCO, Lacan's seminar that year focused on repetition and automatisms as psychoanalytic symptoms. Freud had suggested that patients' semimechanical repetitions revealed trauma. Lacan found in information sciences a rational system for their analysis. From informational Markov chains, cybernetic loops, and game-theoretical matrices he elicited suggestions for defining a semistructured symbolic order bearing down on patients' patterned expressions. Lacan turned Poe's account of a circulating letter into an example of how communication fixes people in intersubjective symbolic chains. It was the subject's place within a series of relays, rather than their ideas or intentions, that mattered. Poe offered perfect material for such analysis: his writings often thematized humans' place within impersonal relations pervaded by technical, mathematical, and scriptural mediations. Poe's passion for cryptography attracted the interest of Shannon (who had based information theory partly on cryptography).[44] Shannon and a Bell Labs colleague, David Hagelbarger, went on to build game-playing automata based on a game described in "The Purloined Letter" in which one player tried to guess if another was thinking of an odd or even number.[45] Shannon playfully termed it the "mind-reading (?) machine," the question mark indicating that its success called the very concept of thought itself into question (figure 5.4). Lacan asked his audience to imagine such a machine, not revealing to them it already existed. Ever the showman, the psychoanalyst provided an in-class demonstration. He handed two of his auditors pencils and paper and exhorted them to quickly write out a series of plus and minus signs, which he announced he would submit to statistical analysis.[46] Due to humans' difficulty generating random numbers, the machines could guess if a human player would choose plus or minus with above average results. By luring people into a series of signs easily processed by machines, it theatrically demonstrated the patterning of human thought. Lacan cited the machine and participants' highly patterned results as evidence of human interactions structured by an impersonal and nonsubjective symbolic order.

Subtending Lacan's invocation of cybernetics was, however, a rejection of identification between human and machine. Take Lacan's indexing of his celebrated "symbolic register," the patterning of speech, to computational processes. "The world of the symbolic is the world of the machine," he argued, with the qualification that "the machine is the structure detached from the activity of the subject."[47] Psychoanalysis, therefore, took as its object the manner in which a human subject located itself within machinic series. Language, as he explained, was not a "code," nor was it "information."[48] "What is redundant as far as information is concerned," that is, the repetitive patterns that could be stripped away from a signal to gain industrial economy, "is precisely what plays the part of resonance in speech. For the function of language is not to inform but to evoke."[49] US-based researchers such as Bateson and Weaver had turned to information theory to model the consistency of mental processes and their obedience to laws of industrial economy and informatic consistency. Lacan used these patterns to highlight a necessary and productive aberrancy: the inconsistency of the psyche. The inability of information theory to assign a stable place to meaning refused human subjects any final or definite positions within its symbolic series. Lacan sought to magnify that indeterminacy,

5.4 Dr. David Hagelbarger of Bell Labs with the SEER, ca. 1952. Source: Image courtesy of the late Dr. Hagelbarger, presumably taken by Bell Telephone Laboratories personnel.

thereby freeing the analysand from the machine-like expressions of trauma that "jammed" their relays.[50]

Static in the Reception

French structuralists' cybernetic efforts instigated conflicts with their American contemporaries. An early sign of the new debate followed Lévi-Strauss's June 1952 address to the *Anthropology Today* conference convened in New York. The event was sponsored by the Wenner-Gren Foundation, a private international foundation based in the United States that promoted anthropology as a tool for avoiding future world wars through the promotion of cross-cultural research.[51] The roster of participants consisted of a number of eminent contributors to the Macy Conferences on Cybernetics, such as Margaret Mead, Kluckhohn, F. S. C. Northrop, and Jakobson. They gathered to discuss essays circulated in advance, including a paper by Lévi-Strauss initially titled "Social Structure." His most programmatic statement on structural methods to date, it posited structuralism as an offshoot of information theory, cybernetics, and game theory, which facilitated the "consolidation of social anthropology, economics, and linguistics into one great field, that of communication."[52] Mead tactfully responded that Lévi-Strauss's version of cybernetics diverged significantly from the account she had developed with her colleagues at the Macy conferences.[53] Northrop objected to modeling all cultures in terms of a single mathematical model derived from Western science.[54] Lévi-Strauss countered that recent advances in communication engineering would allow for experimental definition of cultural pattern, implicitly insulated from provincial cultural prejudice.[55] His explanations did little to assuage his detractors, and the debate itself highlighted an intractable incongruity between Lévi-Strauss's attempts at defining "communications" as a technoscientific enterprise that transcended culture and history, on one hand, and the reality of "communications" as a highly politicized problem inextricable from its situated theorization, on the other.

In France, Lévi-Strauss's synthesis of Mauss and cybernetics elicited opposition from intellectuals whose orientations were more classically French and Marxist. The Cahiers internationaux de sociologie, in particular, became an important forum for resisting structural anthropology as Lévi-Strauss defined it. Chief among the opponents was Georges Gurvitch, who had taught with Lévi-Strauss at the ELHE. In "The Concept of Social Structure" Gurvitch argued that Lévi-Strauss's scientism replaced real social contradictions with ahistorical mathematical symmetry.[56] Essays by Alain Touraine and Henri Lefebvre on American social science and its concept of totality radicalized the critique by introducing a stronger emphasis on the role of class domination in social-scientific reason.[57] Lefebvre in particular would refine this critique into a damning indictment of structuralism. In his 1958 essay "Marxism and the Theory of Information," he ridiculed structuralists' claim that techniques for measuring telegraph transmissions provided suprahistorical procedures for understanding anthropological and sociological arrangements. He dismissively labeled cybernetics and information theory as sciences of "apparatuses that maintain and consolidate a *structure* which has been determined within and by an information machine."[58] In other words, Lefebvre suggested that structuralists ontologized and universalized the artifactual and contingent structures of machines. Among "human scientists" not persuaded by structuralist methods, a brisk trade in literature on technocracy as a method of industrial domination thrived, with notable contributions by Touraine, Nora Mitrani, and Claude Lefort, among others.[59]

Existentialists joined Marxist intellectuals in recoiling from structuralism. Interviewed in 1966, philosopher Simone de Beauvoir claimed that structuralist methods absolved bourgeois consciousness of history, practice, and political commitment. Through its analytics, "there exists neither misery nor unhappiness; there are only systems."[60] Her husband, philosopher Jean-Paul Sartre, told another interviewer, "in a technocratic civilization, there's no more room for philosophy. . . . Look at what's happening in the United States! Philosophy has been replaced by the behavioral sciences [*sciences humaines*]," and thinking replaced by "technician philosophers."[61] These charges amounted to an attack on the research university taking root in postwar France under the influence of US models. Where later Anglophone readers would find in structuralism a new face for Marx, Freud, Heidegger, and Nietzsche, early French readers like Sartre found dogmatic scientism in the thrall of depoliticized US social science.

Cybernetics' Colonial Unconscious

A major theme of the emerging French reception of communication theory was fear over decolonization and imperial decline. Ambivalent allusions to *ensauvagement*—a right-wing French anxiety over becoming "primitive" or "savage"—inflected cybernetic commentaries. Characteristic was literary critic Pierre Rotenberg's 1968 essay "Reading Codes," published by *Tel Quel*, which likened processes of scientific and economic standardization to a "codification" that imposed "internal colonization" on the French.[62] This reading conflated technical modernization, particularly the techno-scientific symbolic manipulation sustained by scientific research, with the political subjugation imposed by French administrators in the colonies. This argument cannily exploited the roots of French human sciences and French technocracy in colonial administration in the nineteenth and early twentieth centuries, to rethink the imperatives driving postwar French modernization. It recognized the boom in the human sciences as a program of governmental and industrial domination turned abruptly inward.

Avant-garde writer Raymond Queneau's introduction to the prospectus for a new encyclopedia in 1956 sketched the rapport of cybernetics to colonialism. Queneau outlines encyclopedic knowledge appropriate to the cybernetic age. He argued that cybernetic feedback, atomic energy, and industrialization in the Global South led "Occidental man to suspect he stands at the dawn of a new era."[63] Advances in information theory and game theory brought about "a true upheaval in the sciences of man."[64] A new encyclopedia informed by these developments would reconsider the place of humankind in the global order. Queneau's upbeat framing of cybernetic encyclopedism hints at French society's tendency to reframe horrific colonial upheaval as an open horizon of technocratic possibility. As historian Sunil Khilnani wrote of postwar French intellectuals, "The critical intellectual, attached to revolutionary principles" seemed to mutate "into a technocratic manager" enamored by systems, codes, and symbols.[65] The rise of the Sixième section and the French social sciences from structural ethnography exemplified this transformation. So-called civilizations "without writing," as Lévi-Strauss termed them, provided foundations for a technical expertise linked to communication science.

De Beauvoir's 1966 novel, *Les belles images* (loosely translatable as "Pretty pictures"), pictured the postwar world as one of repressed technocrats managing planetary desire through communication.[66] The character at the center of the novel, Laurence, is a successful advertising executive who uses marketing research and psychological profiles to deftly adjust public desire.[67] De Beauvoir identified communication professionals with a technocratic ideology promising to deliver wealth and happiness to the Global South. An anonymous narrator explains that synthetic proteins, birth control, automation, and atomic energy would bring about planetary abundance and leisure by 1990, even in "black Africa." "Machine translation and material abundance would transform the world's peoples into a single and well-provided-for entity."[68] When conversation turns to IBM computers said to create art and compose music, the characters agree that "the idea of what constituted man was due to be overhauled and no doubt it would vanish; it was a nineteenth-century invention and now it was out of date."[69] Not only global hunger, but music, social inequality, and desire itself were among the problems to be snuffed out by a technocracy that eclipsed the very idea of man.

No work captured the colonial unconscious of French technocracy better than Lévi-Strauss's 1962 masterpiece, *The Savage Mind*. *The Savage Mind* challenged the centuries-old prejudice that European scientific and technical reasoning was superior to "primitive thought." Lévi-Strauss contended that modern and primitive cultures alike were organized around a complex system of informational "codes" that functioned somewhat like metadata, ordering empirical datum from above. Yet these codification schemes varied across cultures. Modern societies, he contended, reasoned according to a "science of the abstract" based on theoretical, instrumental, and mathematical precision, as exemplified by the figure of an engineer. Primitive groups employed a "science of the concrete" that fashioned meaning from the ordering of materials at hand.

At first glance, *The Savage Mind* appeared to reduce all indigenous cultures to a manifestation of recent informatic findings—preserving, as it were, the epistemic and cultural supremacy of the West. Lévi-Strauss, however, suggested the shoe was on the other foot. In an almost euphoric conclusion, he celebrated the great genius of the savage mind to have long ago recognized and understood what Western information theorists had only recently discovered: that the world is organized into signals for our interpretation. In treating animals, plants, and other aspects of the natural world as a system of obscure signs, the savage mind had discovered "principles of interpretation whose heuristic value and accordance with reality have been revealed to us [Westerners] only recently through very recent inventions: telecommunications, computers, and electron microscopes."[70] Where *Race and History* and *Tristes Tropiques* had identified colonialism with cybernetic entropy, *The Savage Mind* suggested a coming global reconciliation at the level of communication sciences. American anthropologist Clifford Geertz remarked in 1967 of *The Savage Mind*, "What a journey to the heart of darkness could not produce, an immersion in structural linguistics, communication theory, cybernetics, and mathematical logic can."[71]

At a time when the new generation of structuralists was oddly quiescent about the brutal French war against Algerian native rule (1954–62), Lévi-Strauss seemed to suggest a utopian resolution of cultural antinomies through digital media.[72] What *The Savage Mind* actually suggested was an epochal reordering of Western culture in which all its familiar landmarks—science, objectivity, humankind, religion—would undergo transformations through the recognition accorded by processes such as decolonization. His erstwhile partner in ethnography and matrimony, Dina Dreyfus, immediately grasped this implication. "If primitive thought escapes irrationality and emotionality, it is only by succumbing to delirious, machinic thought," she wrote.[73] If so, however, then modern science, too, must find itself transformed by a "primitive thought [that] instructs us on the meaning and validity of cybernetics and information theory."[74] Western science was no longer master of its house, and neither anthropology nor communication science could provide definitive measure of that domain. Lévi-Strauss expressed this in the famous remark at the end of the book, "I believe the ultimate goal of the human sciences to be not to constitute, but to dissolve man."[75] Confrontation with cybernetics, followed by indigenous knowledge that explained its true meaning, demanded "the total overturning of any preconceived idea" of the human sciences and their object. The final lesson of cybernetics was that of an indigenous wisdom that rendered both cybernetics and its anthropologies untenable.

Learning to Code

The Savage Mind occasioned a broader reassessment of the human sciences marked by a new ascent of "coding" as a key concept poised to dislocate, and perhaps dissolve, existing scientific hierarchies. In a nationally televised conversation with philosopher Alain Badiou in 1966, Foucault categorized information theory, cybernetics, and the writings of Lacan and Lévi-Strauss as a single phenomenon characterized by the "reflexive relation of science to itself."[76] Speaking on linguistics and the social sciences at a panel in 1968 in Tunis, Foucault credited Saussurean linguistics with abandoning "language as a translation of thought" in favor of the modern idea of language "as a form of communication" involving senders, receivers, messages, and codes.[77] This Saussurean approach reflected its cybernetic refashioning by the likes of Jakobson and his colleagues at MIT's Research Laboratory of Electronics (RLE). According to Foucault, linguistics now joined all disciplines engaged in "the analysis of all forms of information."[78] Informatic reason also circulated in popular discourse, as thinkers from philosopher Jean Hyppolite to Bourdieu drew on the jargon of information theory to make their disciplines intelligible in televised appearances.[79] Having percolated through the seminars, laboratories, and research centers of Paris, communication theory now brought the human sciences to the general public on the backs of state-funded national communication initiatives.[80]

Learning to code—that is, to cast cultural objects in terms of codes, relays, patterns, and systems—did more than reframe existing knowledge in cybernetic jargon. It also reflected a growing cynicism toward existing cultural and scientific modes. From the 1960s onward, the semiotic task of deciphering obscure "codes" in culture, politics, and science overtook the structuralist project. This crypto-structuralism shifted emphasis from the neutral connotations of "communication" to antagonistic notions of "code."[81] To speak in code is to court obfuscation, misunderstanding, and subterfuge. As Lacan put it, "It is the first time that confusion as such—this tendency there is in communication to cease being a communication, that is to say, of no longer communicating anything at all—appears as a fundamental concept."[82] The writings of Barthes and other affiliates of the journal *Tel Quel* particularly emphasized this connotation, with analytics that contested the neutrality of communication even as they read communicative codes into all manner of phenomena.[83] If these terms furthered the technocratic project of US foundations, they also set in motion a radical critique of scientific neutrality. Beneath the neutral science, something "savage" lurked.

The centers, seminars, and working groups of postwar France formed an infrastructure circulating new informatic codes across the human sciences. These centers' emphasis on team-based collaborative inquiry demanded provisional working languages, which communication theory had provided. The aforementioned CECMAS (figure 5.5), frequented by a diverse cast of linguists, literary critics, sociologists, poets, semioticians, ethnographers, and philosophers, was one such proving ground for communication theory. Scholars converged around this mélange of structural and informatic tropes given shape by Jakobson and Lévi-Strauss. Indeed, the center seemed to realize the visions of Dubarle, Lévi-Strauss, and the Rockefeller Foundation of a center for scientific, experimental, and empirical research. It also reflected the kind of interdisciplinary communication Jakobson had already undertaken at MIT with colleagues at the RLE.[84] The influential journal launched by scholars at CECMAS, *Communications*, provided an instrument for propagating its discourse.

The function of cybernetic discourse at CECMAS mirrored that of the American institutions of communication research CECMAS sought to emulate. Recently founded centers in the United States, such as the Mental Research Institute (tied to Bateson and the Palo Alto Group) and the Institute for Communications Research (ICR, home to Wilbur Schramm, who spearheaded the publication of Shannon and Weaver's *The Mathematical Theory of Communication*), had derived a paradigm for interdisciplinary coordination from cybernetics. The language of information offered an intellectual framework conducive to the pursuit of external funding, thus lending currency to experimental and emerging topics around ill-defined objects. Directors of CECMAS identified Columbia's Bureau of

MINISTÈRE DE L'ÉDUCATION NATIONALE
ÉCOLE PRATIQUE DES HAUTES ÉTUDES
SCIENCES ÉCONOMIQUES ET SOCIALES (6ᵉ section)

Centre d'Études des Communications de Masses
C E C M A S

Paris, le **13 Oct 62**

Cher ami,

J'espère que vous ne nous oubliez pas; vous savez que nous comptons sur vous pour dans contributions à notre n° 2 de Com-munications: 1) un texte, dont vous m'aviez parlé, sur Science et Culture de masse pour notre Débat sur Enseignement et Culture

5.5 The Center for the Study of Mass Communications letterhead noted that the center belonged to the school for "economic and social sciences" and was overseen by the "national ministry of education." This located it within modernization schemes of the period that cultivated social analysis for public and industrial administration as well as for a broader push by the ministry of education to exploit communication infrastructures for mass instruction. Source: Roland Barthes to Jacques Durand, October 13, 1962, original held at the Institut mémoire de l'édition contemporaine, Saint-Germain-la-Blanche-Herbe, France. Digitized and posted online at "Jacques Durand: Un dialogue avec Roland Barthes," http://jacques .durand.pagesperso-orange.fr/Site/Textes/t4.htm.

Applied Social Research (institutional home to Paul Lazarsfeld of the Macy Conferences on Cybernetics), Schramm's ICR, and communication theorist Bernard Berelson (founder of the Center for Advanced Studies in the Behavioral Sciences and leading sponsor of researchers in cybernetics, information theory, AI, and systems theory) as models for the new Parisian center.[85] As one of its members explained, CECMAS was established on the belief that the they should not leave the study of audiovisual communication "all to the Americans."[86] Rather than invent the field from scratch, communication researchers in France would borrow from their peers in the New World. Lazarsfeld had even attended the inaugural planning meeting for CECMAS with Barthes and Edgar and Violette Morin.[87]

The establishment of CECMAS at the late date of 1960 prohibited the neutral application of cybernetic analysis that had thrived in the previous decade. Jakobson and Lévi-Strauss, not to mention Wiener himself, had been touting the importance of cybernetics around Paris for close to a decade. In that time critiques of cybernetics and information theory had been widely disseminated in French intellectual scenes. The skeptical commentaries of Gurvitch, Lefebvre, Canguilhem, and Lacan belied the ostensible neutrality of informatic models, which was soon to be further challenged by the ambivalent conclusions of *The Savage Mind*.[88] This presented a conundrum. The administrative and interdisciplinary tasks confronting the institute, and the intellectual terrain it chose to address, encouraged cybernetic analysis. Insofar as its researchers wanted to join an intellectual discussion with their American contemporaries in communication research, a turn toward the jargon of information theory—already deeply embedded in the structuralist paradigm—plugged their work into an international conversation. Yet the intellectual milieu in which CECMAS thrived questioned the supposed neutrality of cybernetic and informational models. The politics of informatics was no secret fact hidden away in foundation archives but very much a matter of public debate. In such a situation, was it more advisable to reject cybernetic reasoning altogether or to rehabilitate it by distinguishing, as Lévi-Strauss had done, between its scientific uses and ideological abuses?

Researchers at CECMAS chose a third path between rejection and rehabilitation: irony. Perhaps owing to the concentration of literary-minded intellectuals associated with CECMAS, its scholars made the productivity of cybernetic discourse a topic for thematization within their own application of that terminology to the analysis of discourse. The result was an experimental (in both the scientific and artistic senses of the word) mode of writing that theorized communication while ironically thematizing its historical and political conditions. Through coursework, conferences, and the center's journal, *Communications*, researchers intermingled the aspirations of American empirical social science with French structuralism and Marxist critique. The center's membership manifested its mixed loyalties. Friedmann contributed regularly to the Marxist-leaning *Cahiers internationaux de sociologie*, while other members showed stronger inclinations toward new structuralist or semiotic methods. Counted among the latter group were Barthes (then a director of studies at the EPHE) as well as a class of up-and-coming graduate students and young lecturers that included literary critics Kristeva, Gérard Genette, and Tzvetan Todorov; film theorist Christian Metz; and sociologist Jean Baudrillard. Noted lecturers from abroad, among them Umberto Eco, occasionally taught at the center as well. Although divided by disciplinary training, national background, and political allegiance, together they developed a critical synthesis that borrowed from the scientistic aspirations of structuralism, the interpretive vocabularies of cybernetics and information theory, and a sensitivity to the historicity and socially situated productivity of science. A sense of artful panache suffused the members' work, with scholars frequently interweaving performative bravado and analytical rigor.

In its early days the center and its members showed something resembling eccentric faithfulness to cybernetics and information theory. Cybernetic reliance on statistical series and on patterned distributions irreducible to human intention (but demonstrably present in human speech and action) provided an extraordinary tool for characterizing the consistency and agency of language and the arts without recourse to conscience or human intentionality. As Barthes put it in the introduction to *Elements of Semiology*, a primer prepared while he was working at CECMAS, "There is no doubt that the development of mass communications confers particular relevance today upon the vast field of signifying media, just when the success of disciplines such as linguistics, information theory, formal logical and structural anthropology provide semantic analysis."[89] According to Barthes, the development of these fields prompted a "demand for semiology."[90] Yet Barthes, too, rejected the possibility of establishing a universal science, cybernetic or otherwise, that would escape the historical circumstances of its own production. As he deployed the tropes of communication, he also submitted their procedures to ideological and historical critique.

Barthes's vital role at CECMAS and his unique genius for turning structuralism into crypto-structuralism related to the peculiar intellectual route that brought him there. Born in 1915 to a working French family of middle-class means, he never earned the prestigious *agrégation*, a sought-after civil accreditation that followed the doctorate. Like many intellectuals of his generation, he served the French state abroad, including stints in Bucharest as a librarian at the Institut français (1947–48) and cultural attaché for the France legation (1948–49), a teaching assistant in Alexandria (1949–50), and subsequently as a civil servant in the Department of Cultural Relations for the Ministry of Foreign Affairs (1950–52).[91] These positions offered him a critical view on the French state even as he found himself—much as Lévi-Strauss and Foucault did in their stints as cultural attachés—an instrument of its foreign policy. Structuralists' supposed quietism in the face of Algeria and May '68 often seemed to reflect a skepticism of bombastic political stances cultivated by their work as civil servants. Asked in 1967 by literary critic Maurice Blanchot to sign a public letter decrying de Gaulle's fascism, Barthes responded, "The political analysis that it implies does not seem very accurate to me. It makes everything dependent on de Gaulle whereas, it seems to me, it is the opposite accusation that must be made, beginning with the classes, the economy, the state, the technocracy."[92] Marching in the street, mobilizing flags, barricades, and bricks, left the real mechanisms of power unseen and untouched. For the crypto-structuralist, oppression dwelled within the code that enchained the university, commerce, governance, and expertise in the production of an entire social order.

Barthes's career, which proceeded not through appointments in Arts and Letters (the classical umbrella of the humanities in France) but rather through external funding and precarious positions in the human sciences (often attached to interdisciplinary working groups) prepared him to study systems over statements. More precisely, it prepared him to grasp speech as so many statements whose intelligibility lay within the system governing their production. Writing in 1946 on the future of literary studies, he pleaded for the assimilation of social and experimental methods into the study of such consistencies. "From now on, criticism must be able to make certain lists, certain calculations and observations of this order."[93] Admitting the barbarism of statistical methods in its reductions of style to one-dimensional numerical tabulations, he wrote that "nevertheless rhetoric must use statistical information."[94] He identified this with interdisciplinary teamwork to be carried out under the auspices of laboratories, institutes, and research seminars. "The death of Belles Lettres" would permit the birth of a new poetics from the "gradual synthesis of the sciences."[95] In defense of integrating literary studies with experimental sciences, he wrote, "exact knowledge of these [scientific] matters will in itself create a kind of freedom, if it is true that freedom is born the day one recognizes a necessity."[96] In the orderly processes governing writing, and the processes imposing order on text, principles of literary, aesthetic, and poetic autonomy would emerge.

Moreover, Barthes had already articulated, in the 1940s, a rationale for the style of "technical writing" that would descend on literature and the human sciences. With the descent into technical and scientific exactitude as well as into the appropriate inventory of terms and analytical methods, a new space of aesthetic freedom might emerge. This claim made him ill-suited for positions in arts and letters at the *grandes écoles* or even the Sorbonne. It did, however, prepare him for the far-reaching reforms of the 1950s and 1960s that ushered in the social sciences and structural methods.

His first serious academic appointments in France were fellowships in lexicology funded by the French Centre national de la recherche scientifique (CNRS). That work led to his appointment at the Sixième section in 1955 as a research assistant for sociologist Friedmann's laboratory project on "social signs and symbols in human relations."[97] Suddenly Barthes found himself at the heart of a booming postwar social-scientific complex, launched with the largesse of American foundations, at the school Lévi-Strauss had helped found. In 1962 he was promoted to director of studies in the sociology and ethnology cluster at the Sixième section. By the time works such as *The Savage Mind* and *The Order of Things: An Archaeology of the Human Sciences* put the human sciences at the forefront of intellectual debate, Barthes was a veteran in battles over their scope and purpose.

Barthes as Code Breaker

Barthes intuitively yoked technical communication to figures of colonial and technocratic power. The driving conceit of his landmark 1957 work *Mythologies*—namely, that French popular culture formed a mythological system akin to those studied by anthropologists—riffed on fears of colonial subjugation already rife in the postwar French mind. In taking the radio, television, cinema, and newspaper as the sites of these mythologies, moreover, he fulfilled the kind of bivalent analysis that the structuralism of Jakobson and Lévi-Strauss advocated for but did not quite implement. Where structural anthropology explained primitive culture with the aid of communication theory, Barthes used French notions of primitivism to explain modern communication. He accentuated the political thrust of this approach by returning, throughout the text, to colonizer mythologies governing military enforcement, government rhetoric, and popular representation. The government's "official vocabulary of African affairs," he wrote, "has no value as communication, but only as intimidation. . . . In a general way, it is a language which functions essentially as a code, i.e., the words have no relation to their content, or else a contrary one."[98] Barthes's framing mobilized communication theory to highlight the opacity and violence of the state media apparatus. Another analysis, of a black French soldier on the cover of a weekly magazine, examined how popular depictions erased the historical specifics of French imperialism and colonialism (figure 5.6). The harm imparted by codes was not that of secrecy but of a kind of exquisitely rational management by, in de Beauvoir's term, "*belles*

5.6   In turning to French imperial iconology for the analysis of "codes," as in this image famously discussed in his book *Mythologies*, Barthes made concrete and political the interrogation of communication underway in structural anthropology. Source: Cover, *Paris Match* 326, July 1955.

images." Barthes's critique, however, carried a second message: language militates against its instrumentalization, for "there is a moment when the words resist."[99] The precise ordering of language, implemented by the state, harbors a fugitive freedom that escapes determination. It is this basic principle that governs Barthes's semiological endeavor, and with it a tactic deploying technical precision against technocracy.

Barthes's 1961 essay "The Photographic Message," published in the inaugural issue of *Communications*, outlines his more concerted efforts to interrogate communications from the early 1960s onward.[100] In exploring how codes intervene even in an indexical work such as a press photograph, he redirects critical analysis from the media object to the codes governing its signification. Some linguists, notably Algirdas Julien Greimas and Georges Mounin, would come to argue that the model's origins in information theory rendered it ill-adapted to the study of language. Barthes, by contrast, indexed the critical value of communication theory to the field of industrial communication that produced it. "The press photograph is a message," wrote Barthes, defined by "a sender, a transmission channel, and its receiver-milieu."[101] Staff, technicians, editors, and readers worked in concert to form this sender-receiver circuit.[102] In a Shannonian twist, he identified as the channel not only the journal, but more specifically its "complex of concurrent messages," which is to say the ensemble of selections that defines the specificity of one transmission or another. This framing upset the schematic idealism of information theory, however, by drawing attention to how a message and its code are produced. Its concretization of informatic terms according to material and economic processes of production, and the attendant interests that structure their organization, exemplifies the materialist and political turn that crypto-structuralism would take in the coming decade.

Most remarkably, Barthes devised a theoretically innovative account of code as a site of struggle and contest where the freedom of a quasi-autonomous sender did battle with the techno-economic constraints on expression. By arguing that "every code is at once arbitrary and rational," he highlighted both the lack of intrinsic meaning to a particular ordering of traces and the limits of a techno-political rationale imposed on that ordering. As dictated by communication theory, semantics falls outside the scope of the code, but engineering, history, channels, and economy—distinctly political forces—nonetheless govern that code, which conditions signification. "Recourse to a code is thus always an opportunity for man to prove himself, to test himself through a reason and a liberty," he continued. "In this sense, the analysis of codes perhaps allows an easier and surer historical definition of a society than the analysis of its signifieds."[103] Shannon's code, conceived as a techno-economic strategy for transmitting signals, morphed in Barthes's hands into the embodiment of political patterns shaping enunciation. Much as engineers could elicit a proximate definition of the limits and probabilities governing a given communication system, the semiotician could define the limits and probabilities— historical and political in origin—that governed a system of signs. Barthes offered a semiotic account of how history and ideology constituted a code, and that code in turn shaped the relative liberty of the readers, writers, and critics deploying it. Over the course of the next decade this conception of code would travel across CECMAS and mark the writings of Baudrillard, Metz, and other fellow travelers.

This emerging project of code-breaking transvalued both science and criticism in a vicious cycle of bivalent oscillation.[104] The critical study of informational economy invited the critical praise of communicative excess. Barthes's celebrated book *s/z*, the result of a seminar he taught at CECMAS from 1968 to 1969, modeled this possibility. Leveraging Jakobson's passing remark about the presence of "sub-codes" beneath every code, Barthes engaged in a virtuosic reading of how Balzac's novella *Sarrasine* militates against any definitive coding. The authorial sender collapsed before the cacophony of subcodes disrupting its message. Barthes likened the story to a "telephone network gone haywire" and claimed that it reversed the logic of formal sciences such as cybernetics and structural linguistics.[105] In a swipe at the binarism of information theory and structural linguistics, Barthes wrote, "One might call idyllic the communication which unites two partners sheltered from any 'noise' (in the cybernetic sense of the word), linked by a simple destination, a single thread."[106] Against this he posed narrative, a site of irremediably dirty and semantic oscillation. "Thus, in contrast to idyllic communication, to pure communication (which would be, for example, that of the formalized sciences), readerly writing stages a certain 'noise,' it is the writing of noise, of impure communication; but this noise is not confused, massive, unnamable; it is a clear noise made up of connections, not superpositions: it is of a distinct 'cacography.'"[107] While strategically retaining concepts of code, encoding, redundancy, and communication to define a readerly text, Barthes contradistinguished their literary value from the efficiency of communication engineering. Barthes inverted the flight from noise that organized Wiener's and Shannon's endeavors. A readerly, narrative text staged a noise that was no longer confused or erroneous. Cybernetics was reduced to a science of proper and orderly encoding—orthography—while semiotics, in its evolving alliance with Marxism, was a science of the improper and errant code: cacography.

From the Seminar

Barthes's insistence in *s/z* on radicalizing analyses of structuralist techniques until they exploded in polyphony complemented his reassessment of the role of the Sixième section in national modernization. Since the founding of his seminar in 1962, he had deployed the seminars as spaces of discursive experiment. As noted, the seminar form embodied the drive toward a more scientistic and research-oriented modern university aligned with the technocratic ambitions of the French state in the 1950s and 1960s. Its standard emphasis on rational and technical analysis found eager reception from French students whose passion for logical and rational analysis French social scientist Hervé Le Bras once labeled an "obedience to the force of reason that one cannot avoid connecting with the rising interest in technology and technocracy."[108] Barthes did not resist the force of that reason, per se, but he transformed his seminar into a space for its critical confrontation. For him, the seminar presented an opportunity to outline the simultaneous imposition of structure and noise within a cultural text.

Barthes's analysis in *s/z* grew out of his efforts to repurpose the university apparatus for a critical confrontation with the political structures imposed on knowledge. Within this exercise, he situated the seminar as a site for students, teachers, and texts to contest technocratic functionality. He described the seminar as a place for the "production of difference," for the superiority of "body over law, of contract over code, of text over writing," that is, as a site where the strictures of technical assignments gave way to scriptural contest.[109] In the seminar, "no knowledge is transmitted (but a knowledge can be created), no discourse is sustained (but a text is sought)," which he contrasted with the linearity of information theory.[110] Students and teacher committed themselves to processing the fixed series of signs generated by a cultural work while refusing any stable assignment to its relations. A rotating cast of visiting writers, critics, and students, including Kristeva and Philippe Sollers, sociologists Baudrillard and Luc Boltanski, psychoanalysts Jacques-Alain Miller and André Green, semioticians Greimas and Metz, philosopher Catherine Clément, architectural theorist

Françoise Choay, and novelist Georges Perec further confounded the imposition of any master code or expertise on the activity of reading.[111]

In lectures and writings finished during the preparation of *s/z*, Barthes linked code-breaking texts to challenges to technocracy. Writing in *Communications* on the protests of May 1968, he described the student movement as imperiling its own agenda through dreams of "social-technocracy."[112] He warned of "functionalist speech" among students who treated the university as a "social function."[113] If cacography scrambled the codes, functionalism (as illustrated by sociologist Talcott Parsons or even Bateson) ensured orderly relay. For this reason, Barthes charged students with "rediscovering some of the watchwords of a previous technocracy ('adaptation of teaching to society's needs,' 'collectivization of research,' primacy of the 'result,' prestige of the 'interdisciplinary,' 'autonomy,' 'participation,' etc.)."[114] The seminars on *Sarrasine* and the resulting book *s/z* captured Barthes's ongoing response to these events. In class, he described the postwar establishment of the EPHE as a network of eighty laboratories and research centers that embodied both the "triumph and summit of the human sciences."[115] Recent reforms published as *National Education and Participation*, he suggested, embodied a new militancy in the alliance between "the sciences of man and technocratic politics."[116] He traced this operation to initiatives that established working groups and scientific teams, increased specialization, and perhaps most offensive to Barthes, that evicted free auditors from seminars. Under the rhetoric of participation and responsiveness to students, the university enacted the transition "from knowledge [*savoir*] to technology."[117]

Having richly benefited from the professional opportunities of postwar educational reforms, by the late 1960s Barthes railed against a new generation of similar reforms. For decades he had creatively adapted critical practice to the emerging scientific and technical fashions sanctioned by state (and less directly, philanthropic) bodies. Faced with the renewed technocratic spirit of the ministers and students of May '68, he laid new emphasis on an often latent, yet persistent, program within his own research to undo techno-scientific regulation from within its own epistemic factories. "Our seminar plays an oppositional role," he told students, based on rejecting language as "mere instrument" in favor of an analysis of "the revolutionary practice of language or, more precisely, of play [*Jeu*]."[118] He invoked two terms central to 1950s structuralism—Jakobson's language as an instrument of communication and Lévi-Strauss's game theoretical appeals to play—to leverage one against the other not as a flight from structures but as a method for destabilization from within. As he told students, their opposition would not take to the streets but rather, "by necessity, remain within the framework of the institution."[119] He suggested their opposition might take the form of "a politics of language that seeks to abolish participants' roles," such as "the paternal role of the professor"; that refuses to convert seminars into "research 'patrols'"; and, finally, that offers students support in the completion of their theses.[120] Through reading, analysis, and dialogue they would counter technocracy set on transforming the university into an engine of well-programmed communication. With these comments in place, he resumed the close structural analysis of *Sarrasine*, literally and figuratively diagramming the codes, redundancies, binary relations, and characters whose strict ordering he would later, in *s/z*, show to be a "telephone network gone haywire," the hostage of cacographical codes that no author, writer, or institution could completely economize.[121]

The Proliferation of Codes

Barthes's contrarian procedures of decoding participated in a larger reconsideration of structural semiotics in the 1960s in terms of a materialist problematic of machines and inscription. Early structuralists seemed to regard structure as quasi-metaphysical forms embodied in, but irreducible to, their relational enactment in matter. Lévi-Strauss, whose *Tristes Tropiques* treated forces such as settler colonialism as agents of entropic decay, was cagey about the actual location of structural force (as a structuring relation it defied any stable assignment). By the late 1960s, the crypto-structuralists, increasingly under the sway of Maoism, sought to localize and politicize these determinations. "Codes" came to designate the orderly, rational laws that were deployed by science, industry, and capitalism but subject to disruption from within by semiology and poetics. Kristeva, a brilliant young graduate student with an appointment at Lévi-Strauss's laboratory, spearheaded this analytical turn. In a series of structuralist writings that extended into the early 1970s, she argued for the need for semiotics to leverage revolutionary poetics against the strictures of scientific code. She cited Norbert Wiener's research on models as a resource for developing a "science of critique" that would be coextensive with a "critique of science."[122] This analysis made explicit the reflexive task of analysis to deploy scientific systems but also interrogate their imbrication with systems of power and domination. She turned her analysis on the language of structuralism and semiology, highlighting how invocations such as communication and code valorized a political logic that subjected speakers by making them faithfully carry out the relays of senders and an overarching code. The science of language could never fully escape these determinations, but it could incorporate critical, poetic, and political dissonance into its enunciation.

Thematizing the political logics of communication carried risks. One of Lacan's followers, Belgian psychoanalyst Irigaray, was ostracized for her emphasis on the gendering of scientific and philosophical discourse. Her 1966 essay "Linguistic and Specular Communication: Genetic Models and Pathological Models" argued for an essential instability in "communication" stemming from the manner in which the visual and imaginary relations she termed "specularization" confounded the crisp distinction between sender and receiver.[123] Specularization actualized "maximum redundancy," throwing the economic codes of language into disarray.[124] This framing contrasted supposedly redundant and disorderly schizophrenic speech with the orderly ideals of communication theory. This tweak of structural psychoanalysis had far-reaching political implications. Irigaray came to view disturbances in communication as instruments for diagnosing oppressive (and irreducibly gendered) schemas bearing down on the subject. She argued in one text that "in the case of schizophrenia, there is no appropriation of the linguistic code in a speech act by an <I> who wants to send a message to a <you>, a message concerning the world—<he/she/it>—constituted as referent."[125] In the specular suspension of the communicative circuit, an entire network of gendered assignments behind sender and receiver fractured into the gendered determinations of language and its acquisition. The fallout from her critiques included not only expulsion from the Lacanian school but also the nonrenewal of her contract at the University of Vincennes, a postwar university loosely modelled on MIT.[126]

Reflecting in 1985 on her work in the 1960s, Irigaray felt "irritated and amused by the language of science" it inhabited, suspecting that in its aspirations of formal rigor she detected "a moral code of 'good conduct,' political economy of the truth."[127] The scholar known for thematizing masculinity in the language of science had found postwar scientific jargon exercising an uncanny pull on her own writing. Having set out to find the difficulty of the mentally ill in negotiating communicative codes, she finally found that her own writing constrained metacommunicative codes. The question she offered up for her patients turned out to be suitable for the entire crypto-structuralist movement: "How does the subject come back to itself after having exiled itself within a discourse? That is the question of any era."[128] From the early 1960s through the early 1970s, and through major works such as Derrida's *Of Grammatology* (1967) and Serres's *Hermès* series (especially 1969–74), a leading theoretical task of the human sciences was to grapple with—and finally expel—the containments of communication analysis. Gradually, Irigaray came to feel it was not gender but technology that imposed an architectonic constraint on analysis. "Language having become the language of technology, where the automaton is master, it is not always easy for the human subject to recognize its own path through the imperatives and circuits of machines."[129] As a young psychotherapist, Irigaray had inhabited the language of cybernetic psychoanalysis with panache. Years later, however, she questioned if even she could measure its ratios within her thought.

As critiques of code proliferated throughout the 1960s, capitalism gradually displaced colonialism as the backdrop concern. Baudrillard's book *The System of Objects*, based on a seminar he taught at CECMAS, directed its readers' attention to "a cybernetic imaginary mode whose central myth will no longer be that of the absolute interrelatedness of the world."[130] His follow-up 1972 essay "Requiem for the Media" directly attacked communication theory as a vehicle of contemporary oppression and accused Jakobson of propagating it. "This theory is accepted practically everywhere, strengthened by received evidence and a (highly scientific) formalization by one discipline, the semio-linguistics of communication, supported on one side by structural linguistics, by information theory on the other," Baudrillard complained.[131] In this account, theorization became a vehicle for erasing the historical foundations of a particular mode of practice and thereby promoting its ability to enforce new societal norms. "The entire conceptual infrastructure of this theory is ideologically connected with dominant practice, as was and still is that of classical political economy. It is *the* equivalent of this political economy in the field of communication."[132]

There was something mundanely factual about Baudrillard's assertion that the rise of information theory in universities was "ideologically connected with dominant practice." While he likely had no knowledge of the ideological agendas that led the Rockefeller Foundation to shepherd the development of social sciences and cybernetics in France, and certainly knew nothing of the CIA's covert links to Levi-Strauss's cybernetics seminar at UNESCO in the 1950s, Baudrillard astutely discerned that the frameworks of cybernetics, information theory, and game theory, when transposed onto the human sciences, remained rigidly oriented toward mapping out the assumptions of industrial engineering. What ultimately allied his critique with that of the crypto-structuralists, rather than with that of his advisor Henri Lefebvre, was his insistence on communicative immanence: there was no outside to codes. He ironically embraced their status as the only maps of meaning rendering the new economies of global communication intelligible. It was not a "false" model, ill-suited for socio-logical or literary analysis; it was an extant order that could be analyzed only from within. In their alignment of communication theory, social sci-ence, and the politics of knowledge, the crypto-structuralists confronted a technocracy to which they owed their thought and careers. While not reducible to these conditions, their interventions entailed an effort to confront the cybernetic logics central to structuralism, which they gen-eralized to a broader critique of the human sciences and the university as instruments of the state. Where some French Marxists and sociologists had sought to drive cybernetics from the university, the crypto-structuralists seized communication theory as an analytical instrument.

In lieu of a story of opposition or subordination to cybernetic (or American) intellectual imperialism, what emerges from the intellectual itinerary shuttling between US cybernetics and the communication-minded francophone theorists is a tale of mutual appropriation. In this unfolding of thought, the language and institutions of cybernetics played a multifaceted role of affiliation and dissociation. Cybernetics radically disrupted intellectual traditions linked to sources like Mauss, phenom-enology, and existentialism, introducing new paradigms of thought that privileged notions of structure, code, rules, and program—which in turn complemented the move toward technocratic administration in French industry, government, and universities. For example, under the influence of US funding, Mauss's critical rejection of technology and calculation as bases for a holistic theory of society gave way to concepts of communica-tion and code that took technical mediation as a model for social theory. As the social and scientific particularities of the United States conquered global thought of the 1950s and 1960s, structuralists took hold of them to remake French thought.[133] In their effort to establish a theoretical approach to the human sciences, historically and socially specific coordinates faded from view. Crypto-structuralists permitted their agendas to be appropri-ated by intellectual models that they, in turn, reappropriated for their own intellectual aims. As the passionate interventions of Sartre gave way to the cool expertise urged on industry and government in the 1950s, the structuralist language of codes, structures, rules, and programs provided an intellectual handle for grappling with these transformations.

From the highly determined milieu of postwar French modernization, invested with the analytics of Marshall Plan consultants, American foundations, and a rising tide of technocratic expertise, communication-minded theorists found a global audience for their work. Yet it is bereft of that context that these debates would become best known to the rest of the world. In the 1960s and early 1970s, major theorists of structuralism and crypto-structuralism did not cease to emphasize the ineradicable embeddedness of their analysis in the scientific, conceptual, and historical context of its production. This makes the global reception of those writings as abstract and mobile theories, amenable to producing standard and disciplined knowledge, all the more peculiar. The oft-repeated charges by French and American critics that "French theory" was jargon-heavy, abstruse, pseudoscientific, or exclusionary reflected, at least in part, the fallout from the crypto-structuralist struggle with technocracy. For some, such as Lacan and Derrida, the recourse to "difficult" writing was an attack on the call for transparency in communication.[134] As these methods left the seminar and escaped Paris, however, their translation engendered new difficulties. Situated interventions devolved into theoretical formulas. Efforts to carry out a program of scientific and technical analysis, itself rife with ironies and in-jokes, gave rise to new cultures of textocratic expertise. Ease with a French phrase and fluency in insider references, theoretical precision, and analytical exactitude gave rise to a new theoretical class. For non-French readers, at times these translations staged, with oppressive exactitude, a simulacrum of communicative clarity foreign to the original works. Thus, a writing scene preoccupied with the situatedness of scientific, technological, and technocratic communication gave rise to standard apparatuses for stabilizing cultural analytics.

Conclusion

Coding Today

Toward an Analysis of Cultural Analytics

Confined as we are to a structural analysis, we need give only a brief justification of the proposition just advanced, and according to which complex [i.e., modern Western] kinship structures—i.e., not involving the positive determination of the type of preferred spouse—can be explained as the result of the development or combination of elementary structures. A special and more developed study is to be devoted to these complex structures at a later date. **Claude Lévi-Strauss, *The Elementary Structures of Kinship*, 1949**

It's complicated. **Facebook, ca. 2007**

In the 2010s, recent liberal arts graduate Anna Wiener quit her job as a Manhattan literary assistant to join a booming digital start-up in Silicon Valley. Her 2010 memoir of that move, *Uncanny Valley*, is rife with stylish commentaries on the pretensions of tech culture: master-of-the-universe venture capitalists, relentlessly confident programming whizzes, and an office decorated with ornamental typewriters. She finds liberation in the shift from editing authors' manuscripts to tinkering with colleagues' programming code. "When I had edited or vetted manuscripts at the literary agency, I moved primarily on instinct and feeling, with the constant terror that I would ruin someone else's creative work. Code, by contrast, was responsive and uncaring. Like nothing else in my life, when I made a mistake, it let me know immediately."[1] Where the literary manuscript demands correlation between word and intent, code is impersonal and indifferent. The human author disappears from its rule-bound series, replaced by industrial and scientific norms supposedly rooted in the economy of mathematical logic.

A closer look at Wiener's colleagues revealed that many were at home not only in the analysis of programming language but also in the series and signifiers of cultural difference. Technical precision in the parsing of signs extended to the structured patterning of identity. Of colleagues on the "solutions team," she wrote, "they had gone to top-tier private colleges and were fluent in the jargon of media studies and literary theory."[2] In her firm's extensive record of user analytics, she found not zeroes and ones, but a vast database ordered along terms more notable in ethnography, sociology, and gender studies than electrical engineering. "Data could be segmented by anything an app collected—age, weight, income bracket, favorite movies, education, kinds, proclivities—plus some IP-based defaults like country, city, cell phone carrier, device type, and a unique identification code."[3] If liberal arts had taught Wiener to read texts for the racial, ethnic, gendered, and national encoding of signifiers, Silicon Valley equipped her with the resources to economically navigate these determinants at scale. In big data, the notion of impersonal and nonsemantic codes governing signification was not merely a "theory"; it was a practice, realized with the aid of venture capital and Markov-style analysis of textual series.

Wiener's memoir of Silicon Valley and its relentless mobilization of signifiers of postindustrial speculation offers a point of entry to considering the legacies of information theory and French theory today. In it, the supposed break of new, data-driven "cultural analytics" from staid literary and critical theory appears somewhat less definitive. It offers clues to how traditions conceived as parallel and sequential—informatics and semiotics; critical theory and data-driven analysis by Google, Facebook, and Amazon—tend toward the development of transversal, intertwined, and cumulative modes of analysis. In her reconstructions of San Francisco Bay Area digital industry, we find a fitting successor to longer traditions of the theoretical and technical analysis of culture. If the University of California at Berkeley commands a special role in the history of so-called poststructural and semiotic inquiry, including the structural analysis of race, gender, and sexuality, perhaps it is not only the unusual draw held by it and other area universities for lecturers like Claude Lévi-Strauss, Gregory Bateson, Michel Foucault, Donna J. Haraway, Judith Butler, and Friedrich Kittler. Perhaps another kind of affinity links their strange new analyses of culture as systems, codes, and structures with a larger burgeoning culture for the sign-based analysis and production of social order.[4] Perhaps the technocracy and the counterculture, programmers and protestors, and systems theorists and cultural theorists collaborated in a new analytics apposite to the start-up cultures of the Internet era and its fallout. Wiener's account offers clues to this common destiny underwriting information theory and French theory and their collaboration in the construction of a code-based analytics tailored to an era of informatic technocracy.

Of course, in the early days of Bateson and his contemporaries' cybernetic enthusiasm, insurmountable obstacles stood in the way of the large-scale codification of culture. In the era of social media, big data, reality TV, the rise of digitally facilitated education and employment, and pervasive citizen surveillance, those constraints seem to have been withdrawn. Digital cultural analytics premised on pervasive informatic enclosure appear to be an imminently realizable endeavor.[5] As at the height of the cybernetics movement, this seems particularly true where the subjects of enclosure are aberrant— in terms of ethnicity, age, forensics, health, income, citizenship—to the supposedly rights-bearing subjects of the population at large. Access, however, has changed—data is largely available to states and private corporations for analysis, with less priority given to linguists, ethnographers, and other university-based experts. This forms a fabric of changes at once technological, political, and economic. In the "liberal West," social data is often codified by proprietary algorithms and enclosed by confidentiality agreements, even as human subjects remain free to go about their daily affairs. The entirety of social life is available for accounting in terms of codes and subcodes. In a twist on the midcentury projects of anthropologists Mead, Bateson, and Lévi-Strauss, the orderliness of difference, ensuring the adequate profile and analysis of groups orthogonal to the mainstream, remains the site of greater interest to cultural analytics. For some platforms, political or ethnic difference might delineate populations of interest. In a peculiar complementarity, some digital humanities projects scour tens of thousands of books and millions of images for scriptural encodings of gendered, classed, and racialized histories. There, enclosure in a corpus is a first step toward the ostensible goal of inclusion in representation.

Among today's practitioners of computational approaches to cultural analysis, research into predecessors in fields like structuralism or even cybernetics often seems thin. Or, more significantly, researchers' predecessors seem defined too narrowly to grasp the genealogy of their research programs. This omission is striking, given how theories of pattern by the likes of ethnographer Ruth Benedict, Mead, and Bateson shaped the meanings attributed to information theory, cybernetics, digital computing, and systems theory in their early decades. One could be forgiven for thinking that proponents of cultural analytics have been so eager to found their methods on a bedrock of computation that they have obscured the contributions made by literary scholars, anthropologists, and psychotherapists to its mineralogical composition.[6] By the same measure, it is hardly clear that present-day proponents of cultural theory have really taken seriously the full implications of the methodological rigor—the search for technical and expert precision—preached by their predecessors in fields like cultural anthropology and semiotics. It is not uncommon to see latter-day proponents of "theory" present it as simply a critique of scientific and instrumental reason, regardless of the essential role of science and information processing instruments in constructing their theoretical programs. Faced with the unimaginably destructive forces of disease, colonialism, nationalism, and pogroms, midcentury cultural theorists distilled cultural analysis into a kind of acid wash for recoding culture as semiotic chains, remote from the crude chauvinisms of racial and totalitarian reasoning. Was something in their method a bit too refined, and too successful, such that the political and technical apparatus of that maneuver disappeared too completely from the reception of their work?

In a first step toward grappling with the historical intermingling of informatics and cultural theory, the proponents of computational methods in the study of culture might do well to revisit the claims of overcoming "theory" that characterized seminal initiatives of the 2000s.[7] Exemplary in this instance are the remarks of Franco Moretti, who said of his computer-aided "distant reading" that "while recent literary theory was turning for inspiration towards French and German metaphysics, I kept thinking that there was actually much more to be learned from the natural and the social sciences."[8] To this end, Moretti embraced operationalization as the foundation of a new method for the study of human culture.[9] Operationalization, he wrote, "describes the process whereby concepts are transformed into a series of operations—which, in their turn, allow us to measure all sorts of objects."[10] Moretti does not embrace all methods loosely identified with digital humanities (for public historians using digital methods to explore more widely available archives, "enclosure" might prove more relevant), but he sketches a compact description for the meaning of rendering theoretical speculation computationally tractable.

It is hard to imagine an intellectual endeavor more resonant with the spirit of the reform-minded ethnographers of the twentieth century, and their successors in French theory, than Moretti's concept of operationalization.

The dream of accounting for culture in terms of operational chains forms a continuous, if occasionally patchy and potholed, bridge between projects as varied as Benedict's "patterns of culture," Macy conference accounts of textual series, and the rise of structuralism. As psychoanalyst Jacques Lacan put it in 1955, with a somewhat precious allusion to the operational sequences of discourse itself, "the ego, like everything else we have been handling of late in the human sciences, is an op-er-a-tion-al notion."[11] Like Lévi-Strauss and Rockefeller Foundation officer Warren Weaver, Lacan turned to operationality as a manner of critiquing the transcendental self valorized by classical humanism.[12] From cybernetics, information theory, linguistics, and anthropology he elicited a matrix of techniques for classifying the individual self as the link in informatically structured communicative series. In lectures such as those on "The Purloined Letter," he sketched the outlines of an analysis of literature, mental illness, and human behavior as forensics in the style of automata theory.[13] That study, a watershed work in so-called French theory, exemplifies a broader turn toward cybernetics, social science, computation, and operations that propelled structural anthropology and structural linguistics to the forefront of postwar intellectual debate. In France, it laid the groundwork for a new theoretically oriented literary criticism critical of what Butler termed "the human subject, the individual, as the metaphysical locus of agency."[14] If we are today revisiting the possibilities of computer-aided investigation and models drawn from the sciences, it is worthwhile to consider their forerunners in projects like settler-colonial administration and robber baron philanthropy.

There is, however, productivity to the forgetfulness. As alluded to above, not only computational analysts but also quite a few cultural theorists display a certain amnesia when it comes to the relation of digital technology to their own disciplines. The long history of using computing to model innovation in the human sciences disappears from much present-day criticism. At least since World War I, proponents of methodological innovation in Europe and the United States championed the promise of emerging scientific instruments, theories, and data collection methods. These arguments worked for nineteenth-century German philologists eager to overcome the religious superstitions of hermeneutics.[15] Their successes later inspired the Rockefeller Foundation to shower funding on the likes of critic I. A. Richards and Jakobson from the 1930s to the 1945s in hope of overcoming philology. In the twenty-first century, the same arguments justify funding computational methods promising to overcome Richards's close reading and various schools of theory (structural, poststructural, semiotic) indebted to Jakobson. Newly funded projects' promises of epistemic revolution prove less persuasive when read closely alongside the promises and methods of their predecessors. But perhaps no project soared and crashed with the magnificence of the cybernetic apparatus. Mead recalled of the Macy conferences, "We were impressed by the potential usefulness of a language sufficiently sophisticated to be used to solve complex human problems, and sufficiently abstract to make it possible to cross disciplinary boundaries. We thought we would go on to real interdisciplinary research, using this language as a medium. Instead, the whole thing fragmented."[16] Ignorance about those histories may prove useful for mobilizing digital humanities. How many mission statements, funding proposals, and peer-reviewed articles would take on new dimensions when juxtaposed with earlier cultural theorists' plans to encode their findings on ethnography, mythology, schizophrenia, and trade into tractable and easily computable punched cards? But knowledge of these projects may prove useful as well. It may be that continuities among cybernetic schemes for codification provide insight into the epistemic purchase of contemporary digital humanists' "codebooks."[17]

For latter-day cultural theorists, moreover, the temptation to dismiss digital analysis as an imperial exercise foreign to the humanities is strong. In *Reading by Numbers*, digital humanist Katherine Bode offers a rich overview of literary critics including Katie Trumpener, Gayatri Spivak, and Jonathan Arac inventorying the failures of Moretti's statistical studies of literature. Bode glosses, "The most general of these criticisms is that, in reducing aspects of the literary field to data, quantitative approaches provide and privilege a simplistic view of literature, one that fails to understand— or more pointedly, dismisses and violates—such things as aesthetic value and literary complexity."[18] But as Trumpener qualifies, the facile opposition between quantitative and qualitative is confounded by the circulation of statistical bibliography and qualitatively minded categories of metadata such as genre.[19] The story becomes even more complex once fields like ethnography, structuralism, semiotics, and systems theory are acknowledged as part of the literary and humanistic traditions of qualitative research. As this book underscores, practices of objectification, quantification, imperial control, data capture and analysis, and colonial adventuring lie at the foundations of these endeavors, imprinting the methods and struggles that define the rise of qualitative and critical theory as a project of the modern research university.

In fact, since the 2000s boom in the digital humanities, a growing number of digital humanists have called for a reassessment of the relations between theory and quantification. While outliers in qualitative and quantitative orthodoxy abide, this book contributes to an emerging consensus that critical and digital inquiry are more interesting as mutual composites than as opponents. As literary critic and digital humanist Ted Underwood explains, "statistical analysis" and "humanistic theory" as "interpretive traditions are often presented as incompatible. But when both traditions are understood, I believe there are no conflicts between them. On the contrary, they dovetail to produce a more interesting and perplexing world."[20] The interrelations of information theory and French theory add to that analysis by looking more closely at the long-standing intermingling of these traditions, suggesting that their supposed incompatibility, while a recurrent trope in the human sciences, is the maker of a struggle within the broad and catholic tradition of humanistic theory rather than a sign of its uncertain relationship with an adjacent field. The supposition of incompatibility speaks less to enduring methodological alternatives than to ongoing struggles over distributions of interpretive power and credit within a broad theoretical community. This means present struggles over cultural analytics cannot be addressed in terms of two cultures or simplified oppositions between data and theory. Rather, the problem of cultural theory is immanent to digital humanities; conversely, digitization and communication theory structure cultural theory and its notion of signifying systems. In lieu of having a new round of theory wars, the humanities need to come to terms with the place of digital and theoretical analysis as forms of representation in their own right, implicated in struggles over interpretation, authority, and the responsiveness of the humanities to political and technical imperatives.[21] Rather than neutralize or valorize methods, such a history situates them within partial struggles carried on by alliances of mixed methods and motives. Neither theoretical nor digital analysis can emerge from such a recounting altogether self-righteous.

An initial step toward that reconstruction understands cybernetics and its corollaries in the human sciences (family therapy, structuralism, semiology, and French theory) as a relatively nonmonetized, conceptual forerunner of today's digital exercises, from data crunching to cultural analytics. While we often think of today's digital platforms as relying on a series of specific technological advances, or even economic modes, they also rest on a series of conceptual presumptions about social life as a symbolic field of networked relations that, if properly mapped, provide general outlines of future exchanges. It's not for nothing that the anthropological, psychoanalytic, and linguistic wings of the cybernetic apparatus share with Facebook a particular interest in mapping sexual and family relations. Cybernetically minded human scientists diagrammed a virtual field of symbolic relations governing social life, producing mirrors of the dendritic diagrams then conquering early computer science. Structuralists sought to reconstruct the transformations governing these distributions. Their tendency to conceive of all phenomena as reducible to differential binary patterns not only recalls contemporaneous projects in digital media but also suggests a vision of the world as the physical repository for highly structured but infinitely diverse combinatorics. At this moment, a reconsideration of structuralism's dreams for social data is instructive for our present.[22]

As seen in chapter 5, Lacan once remarked that "the world of the symbolic is the world of the machine."[23] This was not because he or his fellow structuralists were influenced by cybernetics per se but because they belonged to a moment in which machinery, records, filing systems, experiments, and, indeed, vast fields of scientific processing had undermined the solidity of bodies, cultures, and history itself. The structuring of the everyday by systems was pervasive, their concretions coming to the surface as petty and profound violence. In the face of genocide and colonialism, Lacan and other cybernetic theorists embraced dynamic networks, data points, and communicative exchange not simply as an ideology of scientism but also as a means of salvation. They turned the dusty philological dictionaries and antiquated ethnographic museums of their predecessors into multimedia archives for cultural simulation. As philosopher Martin Heidegger, the onetime National Socialist apparatchik put it, quoting Hölderlin and perhaps explaining away his own collusion, "But where the danger lies, also grows the saving power."[24] Where structuralists' predecessors traveled the world collecting traces of dead and dying cultures, structuralists sought to reduce these records to a set of dynamic data points for logical and mathematical transformation. Where Mead's and Bateson's predecessors saw primitive cultures in want of scientific analysis, they found perfectly logical systems whose elegance was confirmed by informatics and unconstrained by the deadening weight of industrial mechanics. In time, these trends would be synthesized by the likes of semiologist Roland Barthes and sociologist Jean Baudrillard, who would find in them a vision of twentieth-century technocracy. In the informatic profiles ethnographers fashioned of primitives and analysts imposed on analysands, they would find the outlines of a managerial strategy based on simulation, repetition, and codification, and destined to colonize the industrial and developed world. Mechanical reproducibility was modern. Codification would be postmodern.

What links might we draw between the cybernetic apparatus and subsequent digital technologies? Internalist histories of computing defined by serial inventions have often viewed social-scientific contributions as inessential to the scientific and technical development of digital technologies. Yet the structuralist gambit occupies a privileged place in the archaeology of cultural analytics central to contemporary IT industries, which invites a reassessment of the essence and origin of computing. Reuniting the history of computing with those of the human sciences and political violence, far from obscuring the history of technology, makes the larger course of its development intelligible. In today's social networks we find, in practice, a mapping of distinguishing features governing the genesis, exchange, and dynamism of communication such as Jakobson or Lévi-Strauss could have only dreamt of. The account rendered of what a social network is, in one patent issued to Facebook Inc. in 2019, even recalls the language of high structural anthropology circa 1953. "A social network," Facebook's lawyers explained for bureaucrats at the US Patent Office, "is a social structure made up of entities, such as individuals or organizations, that are connected by one or more types of interdependency or relationships, such as friendship, kinship, common interest, financial exchange, dislike, or relationships of beliefs, knowledge, or prestige."[25] "Network." "Social structure." Not an association of people or things but "entities" defined not by qualities or embodiments but by "relationships," such as "kinship," "exchange," or "beliefs," whose "interdependency" provides a model of future transformations. The diagrammatic account of how such relations take shape around nodes and edges—where the relevant inputs could be cousins, lovers, news items, or recommended vendors—are all issues treated in some detail by Mead, Bateson, and Lévi-Strauss, reflected on by Jakobson, and given an ironic gloss in writings by Barthes, Baudrillard, and other crypto-structuralists.

In Facebook's network diagram, as in Shannon's schematic accounts of communication and genetics, Bush's informatic coding of fingerprints, Jakobson's informational representation analysis of phonemes, Mead's kinship diagrams, Bateson's studies of metacommunication, or the Human Relations Area File (HRAF) populating Lévi-Strauss's laboratory, we have

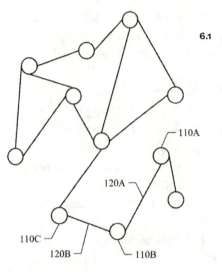

110A

120A

110C

120B

110B

6.1 "Each node may represent an entity, which may be human (e.g., user of the social-networking system) or non-human (e.g., location, event, action, business, object, message, post, image, web page, news feed, etc.)." Source: Patent for applicant Facebook of a work by inventors Michael Johnson, Michael Dudley, and Ryan Patterson, "Suggesting Search Results to Users before Receiving Any Search Query from the Users," United States US10467239B2, filed June 8, 2018, and issued November 5, 2019. https://patents.google .com/patent/US10467239B2/en.

not so much an assignment of empirical givens but a paradigm of storage, correlation, and extrapolation—of database and metadata. To think for a moment in terms familiar to Lévi-Strauss, it schematized the belonging of *les données* to *les dons*, that is, of "data" to the structure that "gives."

Yet schematic formulations of genetics, communication, speech acts, kinship, or mental pathology are not only descriptions of generativity; they are also, themselves, generative. Their synthesis of political, technological, and economic turmoil gave birth to our present. They are a modern answer to the problem of modern violence, an archive aimed at transcending the constraints of historicism. It was grappling with these problems that drove theorists deep into thickets of social relations, networks, and exchanges, emerging from them with new proposals for using datafication against menace. This confrontation with modern problems would, under quite different circumstances, permit Harvard University undergrads' networked rating system of potential mates to scale up, with the speculative resources of Silicon Valley, into the world's highest-valued social network (by investors and markets, that is). If the intersections and parallels between two generations of databasing are clear, these circumstances also point to an epistemic and political gulf. In the case of the cybernetic apparatus, utopian ambitions for vast symbolic inventories and extrapolations realized the conflicting impulses of robber barons' Progressive Era reform initiatives and the subsequent generation of migrant intellectuals' search for a counter-apparatus to the destructive powers of the modern era. Both were exercises in databasing indexed to historical threats. In the present era of mass extinctions, viral exceptions, and the looming specter of untold climatic migrations, are there any imperatives besides monetization and state surveillance to rationalize today's far more comprehensive apparatuses?[26]

Notes

Introduction

1   Margaret Mead, "The Study of Culture at a Distance," in *The Study of Culture at a Distance*, ed. Margaret Mead and Rhoda Métraux (Chicago: University of Chicago Press, 1953), 40.

2   Gregory Bateson, *Naven: A Survey of the Problems Suggested by a Composite Picture of the Culture of a New Guinea Tribe Drawn from Three Points of View* (Stanford, CA: Stanford University Press, 1958 [1936]), vii. Cited comments from the newly written preface for the 1958 edition.

3   Norbert Wiener, *Cybernetics, or Control and Communication in the Animal and the Machine* (Paris: Hermann and Cie, 1948), 38.

4   Warren S. McCulloch, "A Recapitulation of the Theory, with a Forecast of Several Extensions," *Annals of the New York Academy of Sciences* 50, no. 4 (1948): 264, https://doi.org/10.1111/j.1749-6632.1948.tb39856.x.

5   On the determinate situations of conceptual analysis, see Samuel Weber, "The Calculable and the Incalculable: Hölderlin after Kittler," in *Media after Kittler*, ed. Eleni Ikoniadou and Scott Wilson (London: Rowman and Littlefield, 2015), 36.

6   Max Horkheimer, "Traditional and Critical Theory" (1937), in *Critical Theory: Selected Essays*, trans. Matthew J. O'Connell (New York: Continuum, 1972), 188–243.

7   Bruno Latour, "Why Has Critique Run Out of Steam? From Matters of Fact to Matters of Concern," *Critical Inquiry* 30, no. 2 (Winter 2004): 225–48.

8   Roland Barthes, *s/z: An Essay*, trans. Richard Miller (New York: Farrar, Straus and Giroux, 1974 [1970]), 264.

9   Daniel Bell, "Notes on the Post-industrial Society: Part I," *Public Interest*, no. 6 (1967): 28.

10   Paul N. Edwards, *The Closed World: Computers and the Politics of Discourse in Cold War America* (Cambridge, MA: MIT Press, 1996), 3–8; Oliver Belcher, "Sensing, Territory, Population: Computation, Embodied Sensors, and Hamlet Control in the Vietnam War," *Security Dialogue* 50, no. 5 (October 2019): 416–36, https://doi.org/10.1177/0967010619862447.

11   On the informatics of domination, see Donna J. Haraway, "Manifesto for
      Cyborgs: Science, Technology, and Socialist Feminism in the 1980s,"
      Socialist Review, no. 80 (1985): 65–108. On the roots of the informat-
      ics of domination in the human sciences, see Donna J. Haraway, "The
      Biological Enterprise: Sex, Mind, and Profit from Human Engineering
      to Sociobiology," *Radical History Review* 20 (Spring/Summer 1979):
      206–37; and Donna J. Haraway, "Signs of Dominance: From a Physiol-
      ogy to a Cybernetics of Primate Society, C. R. Carpenter, 1930–1970,"
      in *Studies in History of Biology*, ed. William R. Coleman and Camille
      Limoges (Baltimore, MD: Johns Hopkins University Press, 1983),
      129–219.

12   Peter Galison, "The Ontology of the Enemy: Norbert Wiener and the
      Cybernetic Vision," *Critical Inquiry* 21, no. 1 (1994): 228–68; Andrew
      Pickering, "Cyborg History and the World War II Regime," *Perspectives
      on Science* 3, no. 1 (1995): 1–48; Edwards, *The Closed World*.

13   Jennifer S. Light, *From Warfare to Welfare: Defense Intellectuals and Urban
      Problems in Cold War America* (Baltimore, MD: Johns Hopkins University
      Press, 2003).

14   See, for example, Chris Hables Gray, ed., *The Cyborg Handbook* (Lon-
      don: Routledge, 1995); Mara Mills, "Deaf Jam: From Inscription to Re-
      production to Information," *Social Text* 28, no. 102 (Spring 2010): 35–58,
      https://doi.org/10.1215/01642472-2009-059; Jennifer S. Light, "Dis-
      criminating Appraisals: Cartography, Computation, and Access to Fed-
      eral Mortgage Insurance in the 1930s," *Technology and Culture* 52, no. 3
      (2011): 485–522; Rebecca Lemov, *Database of Dreams: The Lost Quest to
      Catalog Humanity* (New Haven, CT: Yale University Press, 2015); Peter
      Sachs Collopy, "The Revolution Will Be Videotaped: Marking a Technol-
      ogy of Consciousness in the Long 1960s" (PhD diss., University of Penn-
      sylvania, 2015); Ginger Nolan, *The Neocolonialism of the Global Village*
      (Minneapolis: University of Minnesota Press, 2018), https://manifold
      .umn.edu/projects/the-neocolonialism-of-the-global-village; William
      Lockett, "The Science of Fun and the War on Poverty," *Grey Room*, no. 74
      (Winter 2019): 6–43; Morgan Ames, *The Charisma Machine: The Life,
      Death, and Legacy of One Laptop per Child* (Cambridge, MA: MIT Press,
      2019); Joanna Radin, "'Digital Natives': How Medical and Indigenous
      Histories Matter for Big Data," *Osiris* 32, no. 1 (2017): 43–64, https://
      doi.org/10.1086/693853; John Durham Peters, "'Memorable Equinox':
      John Lilly, Dolphin Vocals, and the Tape Medium," *boundary 2* 47, no. 4
      (2020): 1–24, https://doi.org/10.1215/01903659-8677814; and Jan
      Müggenburg, "From Learning Machines to Learning Humans: How Cy-
      bernetic Machine Models Inspired Experimental Pedagogies," *History of
      Education* 50, no. 1 (2020): 112–33, https://doi.org/10.1080/0046760X
      .2020.1826054.

15 Lisa Nakamura, speaking on the roundtable "Systemic and Epistemic Racism in the History of Technology Organizer," Society for the History of Technology Annual Meeting, November 19, 2021 (online).

16 In a number of instances, issues of sexual alterity seem also relevant, as in the commentaries of Mead, Barthes, and Foucault that bear on kinship, gender, and institutionalization.

17 N. Katherine Hayles, *How We Became Posthuman: Virtual Bodies in Cybernetics, Literature, and Informatics* (Chicago: University of Chicago Press, 1999); Haraway, "Manifesto for Cyborgs"; Friedrich A. Kittler, *Gramophone, Film, Typewriter*, trans. by Geoffrey Winthrop-Young and Michael Wutz (Stanford, CA: Stanford University Press, 1999 [1986]); Rosi Braidotti, *The Posthuman* (Cambridge: Polity, 2013); Claus Pias, "The Age of Cybernetics," in *Cybernetics: The Macy Conferences 1946–1953; The Complete Transactions*, ed. Claus Pias (Berlin: Diaphanes, 2016), 11–26. On cybernetics ties to the (post)structural critique of humanism, see Céline Lafontaine, *L'Empire cybernétique: Des machines à penser à la pensée machine* (Paris: Seuil, 2004); François Cusset's response "Cybernétique et « théorie française »: Faux alliés, vraies ennemis," *Multitudes Web* 22 (Autumn 2005); Lydia H. Liu, *The Freudian Robot: Digital Media and the Future of the Unconscious* (Chicago: University of Chicago Press, 2010); Warren Sack, "Une Machine à raconter des histoires: De Propp aux software studies," trans. Sophie B. Rollins, *Les Temps Modernes*, no. 676 (November–December 2013); and Jérôme Segal, Le zéro et le un: Histoire de la notion scientifique d'information au XXe siècle (Vol. I) (Paris: Éditions Matériologiques, 2013, 485–532.

18 Pias, "The Age of Cybernetics," 17, 18.

19 Claude Lévi-Strauss, *The Savage Mind* (Chicago: University of Chicago Press, 1966), 247.

20 Hayles, *How We Became Posthuman*, esp. 84–112; Mark B. N. Hansen, *New Philosophy for New Media* (Cambridge, MA: MIT Press, 2004).

21 Leo Marx, *The Machine in the Garden: Technology and the Pastoral Ideal in America* (New York: Oxford University Press, 1964); James W. Carey, "Technology and Ideology: The Case of the Telegraph" (1983), in *Communication as Culture: Essays on Media and Society* (New York: Routledge, 2009), 155–77; Laura Otis, *Networking: Communicating with Bodies and Machines in the Nineteenth Century* (Ann Arbor: University of Michigan Press, 2001); David E. Nye, *American Technological Sublime* (Cambridge, MA: MIT Press, 1994); Jeffrey Sconce, *Haunted Media: Electronic Presence from Telegraphy to Television* (Durham, NC: Duke University Press, 2000).

22 Hall spoke of "American theories and models" that abolished contradiction in favor of "dysfunctions" and "tension management." Stuart Hall, "Cultural Studies and the Centre: Some Problematics and Prob-

lems," in *Culture, Media, Language: Working Papers in Cultural Studies, 1972–79*, ed. Stuart Hall (London: Hutchinson and Centre for Contemporary Cultural Studies, University of Birmingham, 1980), 20.

23 Margaret Mead, "Cybernetics of Cybernetics," in *Purposive Systems: Proceedings of the First Annual Symposium of the American Society for Cybernetics*, ed. Heinz von Foerster et al. (New York: Spartan Books, 1968), 2.

24 Steve J. Heims, *Constructing a Social Science for Postwar America: The Cybernetics Group (1946–1953)* (Cambridge, MA: MIT Press, 1991); Geoffrey Bowker, "How to Be Universal: Some Cybernetic Strategies, 1943–70," *Social Studies of Science* 23, no. 1 (1993): 107–27; Ronald R. Kline, *The Cybernetics Moment: Or Why We Call Our Age the Information Age* (Baltimore, MD: Johns Hopkins University Press, 2015).

25 Kline, *The Cybernetics Moment*, 7.

26 For more on these overlaps see, for example, Ronald R. Kline, "What Is Information Theory a Theory Of? Boundary Work among Scientists in the United States and Britain during the Cold War," in *The History and Heritage of Scientific and Technical Information Systems: Proceedings of the 2002 Conference, Chemical Heritage Foundation*, ed. W. Boyd Rayward and Mary Ellen Bowden (Medford, NJ: Information Today, 2004), 15–28; Claude E. Shannon, review of *Cybernetics, or Control and Communication in the Animal and the Machine*, by Norbert Wiener, *Proceedings of the Institute of Radio Engineers* 37 (1949): 1305; and Kline, *The Cybernetics Moment*, 7–8, 69–83.

27 Claude E. Shannon and Warren Weaver, *The Mathematical Theory of Communication* (Urbana: University of Illinois Press, 1949).

28 Dorothy Ross, *The Origins of American Social Science* (Cambridge: Cambridge University Press, 1991).

29 Harold D. Lasswell and Myres S. McDougal, "Legal Education and Public Policy: Professional Training in the Public Interest," in *The Analysis of Political Behaviour* (London: Kegan Paul, Trench, Trubner, 1947), 33. On Lasswell's communication research supported by the Rockefeller Foundation, see Rockefeller Foundation, *The Rockefeller Foundation Annual Report 1942* (New York: Rockefeller Foundation, 1943), 38–39; and Brett Gary, "Communication Research, the Rockefeller Foundation, and Mobilization for the War on Words, 1938–1944," *Journal of Communication* 46, no. 3 (1996): 124–48. On its postwar development, see Fenwick McKelvey, "The Other Cambridge Analytics: Early 'Artificial Intelligence' in American Political Science," in *The Cultural Life of Machine Learning: An Incursion into Critical AI Studies*, ed. Jonathan Roberge and Michael Castelle (London: Palgrave Macmillan, 2020), https://doi.org/10.1007/978-3-030-56286-1.

30 Lasswell and McDougal, "Legal Education and Public Policy," 33.

31   Jürgen Habermas, "Technology and Science as 'Ideology'" (1968), translated by Jeremy J. Shapiro," in *Toward a Rational Society* (Boston: Beacon, 1970), 81–127.

32   Samuel Weber, *Institution and Interpretation* (Minneapolis: University of Minnesota Press, 1987), 25.

33   Weber, *Institution and Interpretation*, esp. 57.

34   Horkheimer, "Traditional and Critical Theory."

35   Claude Lévi-Strauss, "Structural Analysis in Linguistics and Anthropology" (1945), in *Structural Anthropology* (New York: Basic Books, 1976), 31.

36   D. N. Rodowick, "An Elegy for Theory," *October*, no. 122 (2007): 91.

37   Louis Althusser, "Philosophy and Social Science: Introducing Bourdieu and Passeron" (1963 lecture), trans. Rachel Gomme, *Theory, Culture and Society* 36, no. 7–8 (2019): 6.

38   Jacques Lacan, "Science and Truth, (1965 lecture, 1966 published), in *Écrits: The First Complete Edition in English*, trans. Bruce Fink (New York: W. W. Norton, 2006), 730.

39   For more on the behavioral sciences and cybernetically inflected rationalism, see Paul Erickson , Judy L. Klein, Lorraine Daston, Rebecca Lemov, Thomas Sturm, and Michael D. Gordin, *How Reason Almost Lost Its Mind: The Strange Career of Cold War Rationality* (Chicago: University of Chicago Press, 2013).

Chapter 1: Foundations for Informatics

Epigraph: Warren Weaver to W. F. Loomis, untitled memo, p. 2, November 28, 1950, RF, collection general correspondence, RG 2–1950, series 100, box 476, folder 3192, RAC.

1   Rockefeller Foundation, *The Rockefeller Foundation: Annual Report 1937* (New York: Rockefeller Foundation, 1938), 39. Note the communication funding reported on page 61. See also David Culbert, "The Rockefeller Foundation, the Museum of Modern Art Film Library, and Siegfried Kracauer, 1941," *Historical Journal of Film, Radio and Television* 13, no. 4 (1993): 495–511; Gary, "Communication Research"; William Buxton, "Rockefeller Philanthropy and Communications, 1935–1939," in *The Development of the Social Sciences in the United States and Canada: The Role of Philanthropy*, ed. Theresa Richardson and Donald Fisher (Stamford, CT: Ablex, 1999), 177–92; William Buxton, "From Radio Research to Communication Intelligence: Rockefeller Philanthropy, Communications Specialists, and the American Policy Community," in *Communication Researchers and Policy-Making*, ed. Sandra Braman (Cambridge, MA: MIT Press, 2003), 295–346.

2   Rockefeller Foundation, *The Rockefeller Foundation: A Review for 1936* (New York: Rockefeller Foundation, 1937), 42.

3   Raymond B. Fosdick, *The Story of the Rockefeller Foundation* (New York: Harper, 1952), 245.

4   On cultural analytics, including the diversity of projects aligned with the term, see Lev Manovich, *Cultural Analytics* (Cambridge, MA: MIT Press, 2020), 7–19. In general, I use the term to refer to the use of informatics or computation to render masses of social or cultural phenomena into "content" for computational analysis and theoretical generalization. I include within the term the use of computational discourse (for example, jargon and diagrams) to reorder the analysis of cultural activities such as speech and mental illness.

5   Warren Weaver, "Recent Contributions to the Mathematical Theory of Communication," in *The Mathematical Theory of Communication* (Urbana: University of Illinois Press, 1949), 114.

6   For an overlapping but distinct history of cultural figuration in Shannon's work (and its relation to competing theories of information), see N. Katherine Hayles, *Chaos Bound: Orderly Disorder in Contemporary Literature and Science* (Ithaca, NY: Cornell University Press, 1990), 31–60.

7   Historical literature on Weaver is relatively thin, at least by comparison with his influence. On his work in informatics, see Philip Mirowski, *Machine Dreams: Economics Becomes a Cyborg Science* (Cambridge: Cambridge University Press, 2002), 169–77; W. John Hutchins, "Warren Weaver and the Launching of MT," in *Early Years in Machine Translation: Memoirs and Biographies of Pioneers*, ed. W. John Hutchins (Philadelphia: John Benjamins, 2000), 17–20; Kline, *The Cybernetics Moment*, 19–24, 27–31, 58, 98; Rita Raley, "Machine Translation and Global English," *Yale Journal of Criticism* 16, no. 2 (November 2003): 291–313, https://doi.org/10.1353/yale.2003.0022; Galison, "The Ontology of the Enemy"; Larry Owens, "Vannevar Bush and the Differential Analyzer: The Text and Context of an Early Computer," *Technology and Culture* 27, no. 1 (January 1986): 78–81; Hunter Heyck, *Age of System: Understanding the Development of Modern Social Science* (Baltimore, MD: Johns Hopkins University Press, 2015), 42–46; and Larry Owens, "Mathematicians at War: Warren Weaver and the Applied Mathematics Panel, 1942–1945," in *The History of Modern Mathematics*, ed. David E. Rowe and John McCleary (Boston: Academic Press, 1989), 287–305. On his influence as an administrator, see Robert E. Kohler, "The Management of Science: The Experience of Warren Weaver and the Rockefeller Foundation Programme in Molecular Biology," *Minerva* 14, no. 3 (1976): 279–306, https://doi.org/10.1007/BF01096274; and Lily Kay, *The Molecular Vision of Life: Caltech, the Rockefeller Foundation, and the Rise of the New Biology* (New York: Oxford University Press, 1993).

8   On human freedom, see Warren Weaver, "Statistical Freedom of the Will," *Reviews of Modern Physics* 20, no. 1 (1948): 31–34, https://doi.org /10.1103/RevModPhys.20.31.

9   Light, *From Warfare to Welfare*; Éric Alliez and Maurizio Lazzarato, *Wars and Capital*, trans. Ames Hodges (South Pasadena, CA: Semiotext(e), 2016).

10   Daniel Nemenyi, "What Is an Internet? Norbert Wiener and the Society of Control" (PhD diss., Kingston University, 2019), 5.

11   For more on these themes, see Ross, *Origins of American Social Science*; and Richard Olson, *Scientism and Technocracy in the Twentieth Century: The Legacy of Scientific Management* (Lanham, MD: Lexington Books, 2016).

12   Edward H. Berman, *The Ideology of Philanthropy: The Influence of the Carnegie, Ford, and Rockefeller Foundations on American Foreign Policy* (Albany: SUNY Press, 1983); Mark Dowie, *American Foundations: An Investigative History* (Cambridge, MA: MIT Press, 2001); Kay, *The Molecular Vision of Life*; Giuliana Gemelli, ed., *The "Unacceptables": American Foundations and Refugee Scholars between the Two Wars and After* (Brussels: Peter Lang P.I.E., 2000); Giuliana Gemelli, *The Ford Foundation and Europe* (Brussels: European Interuniversity Press, 1998).

13   Judith Sealander, *Private Wealth and Public Life: Foundation Philanthropy and the Reshaping of American Social Policy from the Progressive Era to the New Deal* (Baltimore, MD: Johns Hopkins University Press, 1997); Fosdick, *The Story of the Rockefeller Foundation*.

14   Fosdick, *The Story of the Rockefeller Foundation*, x.

15   Fosdick, *The Story of the Rockefeller Foundation*, 15.

16   Dowie, *American Foundations*, 27–28, 56–57, 107.

17   On film and broadcasting as instrumentalities see John Marshall to David H. Stevens, January 22, 1936, RF, interoffice correspondence, "Program and Policy—Radio and Motion Pictures 1914–1940," RG 3.1, 911, box 5, folder 50, RAC.

18   On the social sciences as instrumentalities for attainment, see "Discussion of Social Sciences Program and Suggestions for Future Development," memo by Robert T. Crane (of the SSRC), October 27, 1938, RF, folder "Program and Policy—Reports 1938–41," RG 3.1, series 910, box 3, folder 16, RAC. On history as a "technique of operation," see David Stevens, "Report to the Trustees' Committee on Humanities in American Institutions," February 15, 1943, RF, "Program and Policy—Reports 1942–1947," RG 3.1, series 9111, box 2, folder 14, RAC.

19   See "Social Technology," 1927, RF, folder "Program and Policy— Reports—Pro 1–4—1914, 1927," RG 3.1, series 910, box 2, folder 10, RAC.

20   Raymond B. Fosdick, *The Old Savage in the New Civilization* (New York: Doubleday, Doran, 1929), 143.

21 Warren Weaver, "The Benefits from Science—Science and Foundation Program—The Proposed Program," January 27, 1933, p. 6, RF, RG 3.1, series 915, box 1, folder 6, RAC.

22 Weaver, "The Benefits from Science—Science and Foundation Program—The Proposed Program."

23 Steven Shapin and Simon Schaffer, *Leviathan and the Air-Pump: Hobbes, Boyle, and the Experimental Life* (Princeton, NJ: Princeton University Press, 1985).

24 For a critique of this liberal political technique, see John P. McCormick, *Carl Schmitt's Critique of Liberalism: Against Politics as Technology* (New York: Cambridge University Press, 1997); and Carl Schmitt, "The Age of Neutralizations and Depoliticizations," in *The Concept of the Political*, trans. M. Konzett and John P. McCormick (Chicago: University of Chicago Press, 2007), 81–96.

25 On social technology, see, for example, "Social Technology," 1927, RF, in "Program and Policy—Reports—Pro 1–4—1914, 1927," RG3.1, series 910, box 2, folder 10, RAC. Regarding the engineering metaphor, see Fosdick, *Story of the Rockefeller Foundation*, 194.

26 These programs were not always successful. Theodor Adorno's rancorous disputes with Paul Lazarsfeld over the administrative thrust of the Rockefeller-funded Princeton Radio Research Project are an example of the difficulties uniting or bridging diverse methodologies and communities.

27 James W. Carey, "Technology and Ideology," 155–77.

28 A full accounting of the movement is not possible here. Like any brief definition of cybernetics, a short gloss on logical positivism suffers from simplification. One analyst suggests the project is better construed as "(1) a reductionist project in which all the sciences would be logically derivable via bridge-laws from physics; (2) a programme for a unified method which would be followed by all sciences; (3) a project for a unified language of science; and (4) a project that would integrate the different sciences, such that, on any specific problem, all relevant sciences could be called upon—a project for the 'orchestration of the sciences.'" John O'Neill, "In Partial Praise of a Positivist," *Radical Philosophy* 74 (December 1995): 33.

29 Weaver to Morris, February 5, 1935, RF, RG 1.1, series 100, box 35, folder 279, RAC.

30 Rodney Koeneke, *Empires of the Mind: I. A. Richards and Basic English in China, 1929–1979* (Stanford, CA: Stanford University Press, 2004).

31 Warren Weaver, "The Average Reading Vocabulary: An Application of Bayes's Theorem," *American Mathematical Monthly* 27, no. 10 (1920): 347–54, https://doi.org/10.2307/2972549; Claude E. Shannon, "The Mathematical Theory of Communication" (1948), in *The Mathematical Theory of Communication* (Urbana: University of Illinois Press, 1949), 26.

32 Susan Shaw Sailer, "Universalizing Languages: *Finnegans Wake* Meets Basic English," *James Joyce Quarterly* 36, no. 4 (Summer 1999): 853–68.

33 Peter Galison, "The Americanization of Unity," *Daedalus* 127, no. 1 (1998): 45–71.

34 Bowker, "How to Be Universal," 107; Kline, *The Cybernetics Moment*, 185.

35 Owens, "Vannevar Bush," 79–80; Colin Burke, *Information and Secrecy: Vannevar Bush, Ultra, and the Other Memex* (Metuchen, NJ: Scarecrow Press, 1994), 38–39.

36 Burke, *Information and Secrecy*, 40.

37 On applications to molecular biology and econometrics, see Vannevar Bush, "Proposal for the Design of an Improved Differential Analyzer," April 22, 1935, RF, RG 1.1, series 224.D, box 2, folder 22, RAC. The quote comes from a March 1936 letter from Bush to Weaver, cited in David A. Mindell, *Between Human and Machine: Feedback, Control, and Computing Before Cybernetics* (Baltimore, MD: Johns Hopkins University Press, 2002). See also the discussion of this quote and the history of the analyzer in Owens, "Vannevar Bush."

38 Owens, "Vannevar Bush," 66.

39 Claude E. Shannon, "A Symbolic Analysis of Relay and Switching Circuits" (1938), in *Claude Elwood Shannon: Collected Papers*, ed. N. J. A. Sloane and Aaron D. Wyner (Piscataway, NJ: IEEE Press, 1993), 471–96.

40 George Boole, *An Investigation of the Laws of Thought, on Which Are Founded the Mathematical Theories of Logic and Probabilities* (London: Macmillan, 1854).

41 Claude E. Shannon, "Mathematical Theory of the Differential Analyzer," *Journal of Mathematics and Physics* 20, no. 4 (1941): 337–54, https://doi .org/10.1002/sapm1941201337.

42 Burke, *Information and Secrecy*, esp. 43–44, 185.

43 Warren Weaver diary entry, April 12, 1937, RF, RG 12, RAC.

44 Vannevar Bush, "As We May Think," *Atlantic Monthly*, July 1945: 101–8; reprinted as Vannevar Bush, "As We May Think," *Life*, September 1945: 112–14, 116, 121, 123–24.

45 Bush, "As We May Think," *Life*, 116.

46 Bush, "As We May Think," *Life*, 123.

47 Bush, "As We May Think," *Life*, 123–24.

48 Transcribed as Claude E. Shannon, "The Potentialities of Computers" (1953), in *Claude Elwood Shannon: Collected Papers*, ed. N. J. A Sloane and A. D. Wyner (Piscataway, NJ: IEEE Press, 1993), 691–94.

49 Claude E. Shannon, "Computers and Automation: Progress and Promise in the Twentieth Century" (1962 lecture), in *Claude Elwood Shannon: Collected Papers*, ed. N. J. A Sloane and A. D. Wyner (Piscataway, NJ: IEEE Press, 1993), 836.

50   See, for example, E. Colin Cherry, "A History of the Theory of Informa-
      tion," *Proceedings of the IEE Part III: Radio and Communication Engineering*
      98, no. 55 (September 1951): 383, https://doi.org/10.1049/pi-3.1951.0082.

51   Cherry, "History of the Theory of Information," 383–84.

52   Jacques Derrida, *Of Grammatology*, trans. Gayatri Spivak (Baltimore,
      MD: Johns Hopkins University Press, 1976 [1967]); Jacques Derrida, *La
      vie la mort: Séminaire (1975–1976)*, ed. Pascale-Anne Brault and Peggy
      Kamuf (Paris: Éditions du Seuil, 2019), 19–46.

53   Vannevar Bush to Barbara Burks in 1938, quoted in Eugen Chiu, Jocelyn
      Lin, Brok Mcferron, Noshirwan Petigara, and Satwiksai Seshasai, "Math-
      ematical Theory of Claude Shannon: A Study of the Style and Context of
      His Work up to the Genesis of Information Theory" (Cambridge, MA:
      MIT, 2001), 34, https://web.mit.edu/6.933/www/Fall2001/Shannon1.pdf.

54   Letter from Vannevar Bush to Barbara Burks, January 5, 1938, Bush
      Papers, Library of Congress, Washington, DC. Quoted in Jimmy Soni
      and Rob Goodman, *A Mind at Play: How Claude Shannon Invented the
      Information Age* (New York: Simon and Schuster, 2017), as well as in Chiu
      et al., "Mathematical Theory of Claude Shannon," 34.

55   Francis Galton, "Arithmetic Notation of Kinship," *Nature* 28, no. 723
      (September 1883): 435, https://doi.org/10.1038/028435b0; Wilson,
      "Visual Kinship," *History of Anthropology Newsletter* 42 (July 24, 2018),
      https://histanthro.org/clio/visual-kinship/.

56   Quoted in Paul A. Lombardo, "Pioneer's Big Lie," *Albany Law Review* 66
      (2003): 1133.

57   Lombardo, "Pioneer's Big Lie," 1134.

58   Heinz von Foerster, Margaret Mead, and Hans Lukas Teuber, "A Note by
      the Editors" (1952), in *Cybernetics: The Macy-Conferences 1946–1953; The
      Complete Transactions*, ed. Claus Pias (Berlin: Diaphanes, 2016), 347.

59   Frank Fremont-Smith, "Introductory Discussion" (1949 address), in
      Cybernetics: The Macy-Conferences 1946–1953; The Complete Transactions,
      ed. Claus Pias (Berlin: Diaphanes, 2016), 29.

60   Donna J. Haraway, "The High Cost of Information in Post–World War II
      Evolutionary Biology: Ergonomics, Semiotics, and the Sociobiology of
      Communications Systems," *Philosophical Forum* 13, no. 2–3 (1981–1982): 253.

61   Von Foerster, Mead, and Teuber, "A Note by the Editors," 347.

62   Wiener, *Cybernetics*, 19–20, 26, 28, 31.

63   Heims, *Constructing a Social Science*, 14.

64   Heims, *Constructing a Social Science*, 60–61.

65   Mary Catherine Bateson, "Continuities in Insight and Innovation: Toward a
      Biography of Margaret Mead," *American Anthropologist* 82, no. 2 (1980): 274.

66   Heims, *Constructing a Social Science*, 54.

67   Lawrence K. Frank, "The Family as Cultural Agent," *Living* 2, no. 1
      (1940): 18, https://doi.org/10.2307/346505.

68 Pias, *Cybernetics*, 34, 43, 64 (remarks spoken in 1949).

69 Heims, *Constructing a Social Science*, 139.

70 Kline, *The Cybernetics Moment*.

71 Wiener, *Cybernetics*; Galison, "The Ontology of the Enemy."

72 Karl Wildes and Nilo Lindgren, *A Century of Electrical Engineering and Computer Science at MIT, 1882–1982* (Cambridge, MA: MIT Press, 1985), 184.

73 Owens, "Mathematicians at War."

74 See Heims, *Constructing a Social Science*, 2–13; Galison, "The Ontology of the Enemy"; and Edwards, *The Closed World*.

75 Jürgen Habermas's phrase inspired by the Palo Alto Group, in *The Theory of Communicative Action, v. I: Reason and the Rationalization of Society*, trans Thomas McCarthy (Boston: Beacon Press, 1984 [1981]): 332, 445n87.

76 On Frank's influence and place in their life, see Bateson, "Continuities in Insight," 273–75; and Lipset, Gregory Bateson, 161–62. On the broader agendas of Frank, see Dennis Bryson, "Lawrence K. Frank, Knowledge, and the Production of the 'Social,'" *Poetics Today* 19, No. 3 (Autumn, 1998), esp., 402, 417–21.

77 Lipset, *Gregory Bateson*, 160–83; David Price, *Anthropological Intelligence: The Deployment and Neglect of American Anthropology in the Second World War* (Durham: Duke University Press, 2008): 19–21, 40–41, 239–43. Frederic Ponten, "Collaborating with the Enemy: Wartime Analyses of Nazi Germany" (PhD diss., Princeton University, 2017), 183–256.

78 Gary, "Communication Research."

79 "Resolved RF 47009," January 16, 1947, p. 47023. RF, RG 1.1, series 224, box 1, folder 2 (MIT, Mathematical Biology, 1945–6), RAC.

80 "Resolved RF 47009," January 16, 1947, p. 47023.

81 "Resolved RF 47009," January 16, 1947, p. 47024.

82 "Resolved RF 47009," January 16, 1947, p. 47023.

83 Frank Fremont-Smith, "Josiah Macy Jr. Foundation Conference Program" (1952), in *Cybernetics: The Macy-Conferences 1946–1953; The Complete Transactions*, ed. Claus Pias (Berlin: Diaphanes, 2016), 531.

84 Von Foerster, Mead, and Teuber, "A Note by the Editors,"342.

85 Norbert Wiener, "What Is Information Theory?," *IRE Transactions on Information Theory* 2, no. 2 (June 1956): 48.

86 Weaver, "Recent Contributions," 95.

87 Weaver, "Recent Contributions," 95 (quote) and 97.

88 Weaver, "Recent Contributions," 116.

89 On the wider cultural context of machine translation, including its intersection with literature and literary criticism, see Sean Michael DiLeonardi, "Cryptographic Reading: Machine Translation, the New Criticism, and Nabokov's *Pnin*," *Post45*, January 2019, https://post45.org/2019/01/cryptographic-reading-machine-translation-the-new-criticism-and-nabokovs-pnin/.

90 Weaver to Wiener, March 4, 1947, box 5, folder 76, NWP. Also available online at http://www.mt-archive.info/Weaver-1949.pdf.

91 Warren Weaver, "Translation," in *Machine Translation of Languages: Four-teen Essays*, ed. W. N. Locke and A. D. Booth (Cambridge, MA: Technology Press of the Massachusetts of Technology and Wiley, 1955), 15.

92 Rockefeller Foundation, *The Rockefeller Foundation: Annual Report 1948* (New York: Rockefeller Foundation, 1949), 20.

Chapter 2: Pattern Recognition

1 The Palo Alto Group's cybernetic definitions of schizophrenia shaped generations of latter-day accounts of electronic media as schizophrenic. For more, see John Durham Peters, "Broadcasting and Schizophrenia," *Media, Culture and Society* 32, no. 1 (January 2010): 123–40; and Jeffrey Sconce, *The Technical Delusion: Electronics, Power, Insanity* (Durham, NC: Duke University Press, 2019).

2 On cybernetics in family therapy, see Deborah Weinstein, *The Pathological Family: Postwar America and the Rise of Family Therapy* (Ithaca, NY: Cornell University Press, 2013), 47–81. On the influence of cybernetics, information theory, and communication theory in the Palo Alto Group more generally, see Carol Wilder, "The Palo Alto Group: Difficulties and Directions of the Interactional View for Human Communication Research," *Communication Research* 5, no. 2 (Winter 1979): 171–86. Members of the Palo Alto Group founded the methods of brief therapy in the late 1960s. For a seminal text on its establishment, see John H. Weakland, Richard Fisch, Paul Watzlawick, and Arthur M. Bodin, "Brief Therapy: Focused Problem Resolution," *Family Process* 13, no. 2 (June 1974): 141–68. For indications of the links between cognitive behavioral therapy and brief therapy, see Lata K. McGinn and William C. Sanderson, "What Allows Cognitive Behavioral Therapy to Be Brief: Overview, Efficacy, and Crucial Factors Facilitating Brief Treatment," *Clinical Psychology: Science and Practice* 8, no. 1 (Spring 2001): 23–37. For an excellent history of the Palo Alto Group by a core member, see Jay Haley, "Development of a Theory: A History of a Research Project," in *Double Bind: The Foundation of the Communicational Approach to the Family*, ed. C. E. Sluski and D. C. Ransom (New York: Grune and Stratton, 1976), 59–104; and John H. Weakland, "One Thing Leads to Another," in *Rigor and Imagination: Essays from the Legacy of Gregory Bateson*, ed. Carol Wilder and John Weakland (New York: Praeger, 1982), 43–64.

3 See, for example, Habermas, *Theory of Communicative Action*, 332, 445n87. and Gilles Deleuze and Félix Guattari, *Anti-Oedipus: Capitalism and Schizophrenia*, trans. Robert Hurley (New York: Viking, 1977).

4   There is a rich body of work attentive to the great open and dynamic possibilities Bateson and his colleagues introduced to the disciplines and popular US thought. For a work attentive to the great promise of Bateson's and Mead's work, while also attentive to larger dilemmas of US postwar society, see Fred Turner, *The Democratic Surround: Multimedia and American Liberalism from World War II to the Psychedelic Sixties* (Chicago: University of Chicago Press, 2013).

5   Ira Jacknis, "Margaret Mead and Gregory Bateson in Bali: Their Use of Photography and Film," *Cultural Anthropology* 3, no. 2 (May 1988): 160–77; Fatimah Tobing Rony, "The Photogenic Cannot Be Tamed: Margaret Mead and Gregory Bateson's *Trance and Dance in Bali*," *Discourse* 28, no. 1 (Winter 2006): 5–27.

6   Friedrich Kittler, "Flechsig/Schreber/Freud: An Informations Network of 1910" (1984), trans. Laurence Rickels, Avital Ronell, David Levin, Adam Bresnick, Judith Ramme, and Friedrich Kittler, *Qui Parle* 2, no. 1 (Spring 1988): 9.

7   Heims, *Constructing a Social Science*, 14–17; Kline, *The Cybernetics Moment*, 40.

8   Wiener, *Cybernetics*, 173.

9   Wiener, *Cybernetics*, 174.

10   John von Neumann, *The Computer and the Brain* (New Haven, CT: Yale University Press, 2000 [1958]).

11   On mechanical standardization in science, see Lorraine Daston and Peter Galison, "The Image of Objectivity," *Representations* 40 (Autumn 1992): 81–128; and Mary Ann Doane, "Temporality, Storage, Legibility: Freud, Marey, and the Cinema," *Critical Inquiry* 22, no. 2 (January 1996): 313–43, https://doi.org/10.2307/1343974.

12   Margaret Mead, *Coming of Age in Samoa: A Psychological Study of Primitive Youth for Western Civilization* (New York: William Morrow, 1928), 8.

13   Margaret Mead, "The Training of the Cultural Anthropologist," *American Anthropologist* 54, no. 3 (1952): 345, https://doi.org/10.1525/aa.1952.54.3.02a00060.

14   Frank F. Tallman, "Organization and Function of the Children's Group of Rockland State Hospital," *Psychiatric Quarterly* 12, no. 1 (March 1938): esp. 187–188, https://doi.org/10.1007/BF01561923; Arthur W. Pense and Alfred M. Stanley, "The Institutional Care of Mentally Ill Children," *Psychiatric Services* 6, no. 9 (September 1955): 5–7, https://doi.org/10.1176/ps.6.9.5.

15   Pesi R. Masani, *Norbert Wiener: 1894–1964* (Basel: Birkhäuser Verlag, 1990), 218; Peter Asaro, "McCulloch, Warren S. (Artificial Neural Networks)," Peter Asaro Ph.D., accessed December 3, 2020, https://peterasaro.org/writing/McCulloch.html.

16   Nathan S. Kline and Manfred Clynes, "Drugs, Space, and Cybernetics: Evolution to Cyborgs," in *Psychophysical Aspects of Space Flight*, ed. B. E. Flaherty (New York: Columbia University Press, 1961), 345–71; Ron R. "Where Are the Cyborgs in Cybernetics?" *Social Studies of Science* 39, no. 3 (June 2009): 39–45.

17   Seth Barry Watter, "Interaction Chronograph: The Administration of Equilibrium." *Grey Room* 79 (Spring 2020): 64.

18   Ruth Benedict, *Patterns of Culture* (1934; repr., Boston: Houghton Mifflin, 1959), 2–3.

19   Jane Howard, *Margaret Mead: A Life* (New York: Ballantine Books, 1980), 175; Rhoda Métraux, "The Study of Culture at a Distance: A Prototype," *American Anthropologist* 82, no. 2 (1980): 364.

20   Gregory Bateson and Margaret Mead, *Balinese Character: A Photographic Analysis* (New York: New York Academy of Sciences, 1942), xvi.

21   Bateson and Mead, *Balinese Character*, xvi.

22   Gregory Bateson and Margaret Mead, "Introduction," in Bateson and Mead, *Balinese Character*, xi. This passage originally appeared inBateson's *Naven* and then reappeared in this introduction, quoted.

23   Gregory Bateson, "Notes on the Photographs and Captions," in Bateson and Mead, *Balinese Character*, 49.

24   Bateson and Mead, *Balinese Character*, xii.

25   Rony, "The Photogenic Cannot Be Tamed," 10. On the long history of colonial administration and anthropology in Bali, see Henk Schulte Nordholt, "The Making of Traditional Bali: Colonial Ethnography and Bureaucratic Reproduction," *History and Anthropology* 8, nos. 1–4: 89–127, https://doi.org/10.1080/02757206.1994.9960859.

26   Bateson and Mead, "Introduction," xiii.

27   Rony, "The Photogenic Cannot Be Tamed," 10–11.

28   Rony, "The Photogenic Cannot Be Tamed," 10.

29   Gregory Bateson, "Social Planning and the Concept of Deutero-Learning" (1942), in George Bateson, *Steps to an Ecology of Mind* (New York: Ballantine Books, 1972), 162.

30   Quoted in Lipset, *Gregory Bateson*, 174.

31   Quoted in Lipset, *Gregory Bateson*, 174.

32   From Gregory Bateson, "Your Memo No. 53," November 15, 1944, quoted in Price, "Gregory Bateson and the oss," 381.

33   In Price, "Gregory Bateson and the oss," 381.

34   Mead and Métraux, *Study of Culture*.

35   Another significant aspect of Bateson's work in this time including German propaganda research. For more on this strand of his wartime communication research, see Ponten, "Collaborating with the Enemy," 183–256.

36   Partial documentation of the exhibit, including links to press releases, can be found online at Museum of Modern Art, *Bali, Background for War: The Human Problem of Reoccupation*, Museum of Modern Art, New York, accessed June 24, 2021, https://www.moma.org/calendar/exhibitions/3141; see, in particular, "Museum of Modern Art Opens Exhibition of Bali, Background for War," press release, August 1943, Museum of Modern Art, New York, https://www.moma.org/documents/moma_press-release_325409.pdf.

37   Gregory Bateson, "Atoms, Nations, and Cultures," *International House Quarterly* 11 (1947): 49, 50.

38   Bateson, "Atoms, Nations, and Cultures," 50. An interesting complement to this approach was Mead's belief in the same period that cross-cultural communication on "the United Nations model" should be based on two world languages, one from a global power (such as English) and another coming from a small, isolated group of people. The idea was to engineer two global worldviews, shaped by the linguistic affordances specific to these differently scaled cultures. Mary Catherine Bateson, *With a Daughter's Eye: A Memoir of Margaret Mead and Gregory Bateson* (New York: William Morrow, 1984), 85–86.

39   Margaret Mead, *Soviet Attitudes toward Authority: An Interdisciplinary Approach to Problems of Soviet Character* (Santa Monica: The RAND Corporation, 1951).

40   The major treatment of these enclosures' cybernetic logic is Edwards, *The Closed World*. See also Beatriz Colomina, "Enclosed by Images: The Eameses' Multimedia Architecture," *Grey Room*, no. 2 (Winter 2001): 7–29, https://doi.org/10.2307/1262540; Jason Pribilsky, "The Will to Enclose: Foucault's Archive in the Era of Cold War Big Data," *Le Foucauldien* 2, no. 1 (2016); and Pamela M. Lee, *Think Tank Aesthetics: Midcentury Modernism, the Cold War, and the Neoliberal Present* (Cambridge, MA: MIT Press, 2020).

41   On social-scientific interest in the family and fascism, see Fred Turner, *The Democratic Surround*, 2 51–53.

42   Ellen Herman, *The Romance of American Psychology: Political Culture in the Age of Experts* (Berkeley: University of California Press, 1995), 240.

43   Herman, *Romance of American Psychology*, 242–43.

44   See Marvin Karno and Donald A. Schwartz, *Community Mental Health: Reflections and Explorations* (Flushing, NY: Spectrum Publications, 1974): 22–23; and Joseph Halpern, *The Myths of Deinstitutionalization: Policies for the Mentally Disabled* (Boulder, CO: Westview, 1980) 1–5.

45   Albert Maisel, "Bedlam 1946," *Life*, May 6, 1946: 102–18; Albert Deutsch, *The Shame of the States* (New York: Harcourt, Brace, 1948).

46   See, for example, Alfred H. Stanton and Morris S. Schwartz, *The Mental Hospital: A Study of Institutional Participation in Psychiatric Illness and Treatment* (New York: Basic Books, 1954); and Erving Goffman, *Asy-*

lums: Essays on the Social Situation of Mental Patients and Other Inmates (Garden City, NJ: Anchor Books, 1961).

47   Margaret Mead, *Male and Female: A Study of the Sexes in a Changing World* (New York: William Morrow, 1967), 251.

48   Mead, *Male and Female*, 251.

49   Lipset, *Gregory Bateson*, 175–76.

50   Lipset, *Gregory Bateson*, 178.

51   On Kees's work with Bateson and Ruesch, see James Reidel, *Vanished Act: The Life and Art of Weldon Kees* (Lincoln: University of Nebraska Press, 2003), 239–348. Unless otherwise noted, I have drawn my accounts of their work together from Reidel's text.

52   Documentation of Bateson's medical films is sketchy, at best, with many works lost. A history of the Palo Alto Group written by Haley lists the films prepared by Bateson, Kees, and Ruesch as the primary data gathered in the first year of the project. "The data that first year were diverse and included the following: a study of otters playing, a study of training of guide dogs for the blind, an analysis of a popular moving picture, a filming of Mongoloid children in a group, analysis of humor and a ventriloquist and puppet, and the utterance of a schizophrenic patient when a project member began to interview him early that year." See Haley, "Development of a Theory," 62. Footage of the otters and dog training films, the patient/subject known as Doris at home, ethnographic films of other families, and footage of other settings and people (including psychoanalyst John Rosen working with patients) are held at the Don D. Jackson Archive, University of Louisiana at Monroe (hereafter JA). The canonical film of *Doris, GB-SU-005* is held in the University of Chicago Digital Media Archive. Additional (perhaps alternate) copies may be in the Human Studies Film Archives at the Smithsonian. Final versions of *Communication and Interaction in Three Families* and *The Nature of Play: Part 1, River Otters* may be found in the Bateson archives at the University of California, Santa Cruz. I received a copy of *Hand-Mouth Coordination* from James Reidel via Henning Engelke as well as a copy of *Approaches and Leavetakings* by Ruesch and Kees. A brief (perhaps incomplete) listing of the films completed by Bateson, Ruesch, and Kees appears in Jurgen Ruesch and Weldon Kees, *Nonverbal Communication* (Berkeley: University of California Press, 1956), 200. The films of "Mongoloid" children appear to be lost.

53   Gregory Bateson, *Steps to an Ecology of Mind* (New York: Ballantine, 1972), x, 179. See also Bernard Dionysius Geoghegan, "Editor's Introduction: What Bound the Double Bind?," *Grey Room*, no. 66 (Winter 2017): 102–9, https://doi.org/10.1162/GREY_a_00213; and Gregory Bateson and Weldon Kees, "The Nature of Play: Part 1, River Otters" (ca. 1954), *Grey Room*, no. 66 (Winter 2017): 110–14, https://doi.org/10.1162/GREY_a_00215.

54   Jurgen Ruesch, "Values, Communication, and Culture: An Introduction," in *Communication: The Social Matrix of Psychiatry* (New York: W. W. Norton, 1951), 6.

55   For foundational analyses of the Bateson films, see Henning Engelke, "Filmisches Wissen und der Geist des Kalten Krieges: Kybernetische Modelle bei Gregory Bateson und Weldon Kees," in *Wissensraum Film*, ed. Irina Gradinari, Dorit Müller, and Johannes Pause (Wiesbaden: Reichert Verlag, 2014), 225–41; and Minette Hillyer, "Camera Documents Made at Home: Visual Culture and the Question of America," *Film History* 27, no. 4 (2015): 46–75.

56   *Communication and Interaction in Three Families*, directed by Jurgen Ruesch, Gregory Bateson, and Weldon Kees (Kinesis Inc.: 1952), 16 mm sound film, approx. 75 min.

57   Ruesch and Kees, *Nonverbal Communication*, 11.

58   Bateson, quoted in Stewart Brand, *II Cybernetic Frontiers* (New York: Random House, 1974), 29.

59   *Hand-Mouth Coordination*, directed by Gregory Bateson and Weldon Kees, 1951, 16 mm, black and white, approx. 10 min. Courtesy the personal collection of James Reidel, furnished with assistance from Henning Engelke.

60   This reflexive turn, which is cybernetic in character, anticipates the rise of video therapy (and video art) and its own mechanisms of feedback. See Peter Sachs Collopy, "The Revolution Will Be Videotaped" (PhD diss., University of Pennsylvania, 2015). This also reflects a broader trend in film from the period of drawing the psychiatrist into the scene of illness, as if by way of implication. See W. J. T. Mitchell, *Seeing Madness, Insanity, Media, and Visual Culture* (Berlin: Hatje Cantz Verlag, 2012).

61   Reidel, *Vanished Act*, 240.

62   Raymond Williams, *Television: Technology and Cultural Form* (New York: Routledge, 2003).

63   Gregory Bateson, "Epidemiology of a Schizophrenia" (1955), in Bateson, *Steps to an Ecology of Mind*, 198.

64   Bateson, "Epidemiology of a Schizophrenia," 198.

65   Bateson, "Epidemiology of a Schizophrenia," 199.

66   Bateson, "Epidemiology of a Schizophrenia," 198.

67   Partial transcript and notes available as Kantor (presumably Jacob Robert Kantor), Henry Brosin et al., "Film" (transcription with session notes), July 28, 1956, in box 18, RBP. Film of the therapy session is archived as "Reels 1–4," Eastern Pennsylvania Psychiatric Institute (EPPI) films Bateson Collection, JA.

68   Resolved RF 52110, "Stanford University—Research in Communication and Psychiatry—Bateson," June 20, 1952, p. 52383, RF, RG1.1 (Project Files), series 200, box 60, folder 718 (FA 386), Stanford University—Bateson, Gregory—(Communications and Psychiatry), RAC.

69 Chester Barnard to J. E. Sterling, April 30, 1952, RF, RG1.1 (Project Files), series 200, box 60, folder 718 (FA 386), Stanford University—Bateson, Gregory—(Communications and Psychiatry), RAC.

70 Chadbourne Gilpatric, "Proposed Grant to Stanford University for Gregory Bateson's Study of Paradoxes of Abstraction," Interoffice correspondence, May 12, 1952, RF, RG1.1 (Project Files), series 200, box 60, folder 718 (FA 386), Stanford University—Bateson, Gregory—(Communications and Psychiatry), RAC.

71 Gilpatric, "Proposed Grant to Stanford University for Gregory Bateson's Study of Paradoxes of Abstraction," Interoffice correspondence, May 12, 1952.

72 Gregory Bateson to Chester I. Barnard, Letter, February 28, 1952, p. 2, RF, RG1.1 (Project Files), series 200, box 60, folder 718 (FA 386), Stanford University—Bateson, Gregory—(Communications and Psychiatry), RAC.

73 Details in "Inventory the Gregory Bateson Papers" (MS 98), Special Collections and Archives, University of California, Santa Cruz, 2008: 4. http://oac.cdlib.org/findaid/ark:/13030/kt029029gz.

74 Personnel listed on page one of Gregory Bateson (presumably), "The Role of the Paradoxes of Abstraction in Communication," interim report and proposal to the Rockefeller Foundation, ca. April 1954, RF, RG1.1 (Project Files), series 200, box 60, folder 718 (FA 386), Stanford University— Bateson, Gregory—(Communications and Psychiatry), RAC.

75 Gregory Bateson, Don D. Jackson, Jay Haley, and John Weakland, "Toward a Theory of Schizophrenia," *Behavioral Science* 1, no. 4 (1956): 251–64.

76 Gregory Bateson, "The Role of the Paradoxes of Abstraction in Communication," RF, RG1.1 (Project Files), series 200, box 60, folder 718 (FA 386), Stanford University—Bateson, Gregory—(Communications and Psychiatry), RAC.

77 Haley, "Development of a Theory."

78 Lipset, *Gregory Bateson*, 215–38.

79 Seb Franklin remarks on the difficulty peculiar to Bateson's work of acknowledging the structures of domination that form the basis of its analyses (as well the larger context of racialized social science to which it belonged). See Seb Franklin, "Commodities, Love, and Attenuated Reproduction in the Archives of Cybernetics," *Diacritics* (blog), June 3, 2021, https://www.diacriticsjournal.com/unarchive-6/.

80 Although initially devised for interviewing families with a schizophrenic member, it was later adapted as a therapeutic tool for general use. See Weinstein, *Pathological Family*, 67–68. For an account of its general use, see Paul Watzlawick, "A Structured Family Interview," *Family Process* 5, no. 2 (September 1966): 256–71. The phrase "scanning for patterns" comes from Watzlawick's article.

81 I am surmising the presence of the fireplace based on my examination of films, but I have not been able to confirm this detail with participants.

82   Watzlawick, "Structured Family Interview."

83   Watzlawick, "Structured Family Interview."

84   Don D. Jackson, "The Eternal Triangle: An Interview with Don D. Jackson, M.D.," in *Techniques of Family Therapy*, ed. Jay Haley and L. Hoffman (New York: Basic Books, 1967), 175–76. Other details related in this paragraph are drawn from Jay Haley, "Family Experiments: A New Type of Experimentation," *Family Process* 1, no. 2 (1962): 265–93.

85   Haley, "Family Experiments," 267.

86   Scenes of this performance may be found in Samuel Moffat, "Area Psychiatrists Study New Schizophrenia Theory," *Palo Alto Times*, May 7, 1958. Hard copies of this text seem hard to find, but Wendel A. Ray at the JA kindly shared a copy with me. For more, see also EPPI films, Bateson Collection, JA.

87   Weakland et al., "Brief Therapy: Focused Problem Resolution," 144.

88   Salvador Minuchin, quoted in Janet Malcolm, "The One-Way Mirror," *New Yorker*, May 15, 1978, 83.

89   Minuchin quoted in Malcolm, "The One-Way Mirror," 83.

90   Paul R. Good, "Brief Therapy in the Age of Reagapeutics," *American Journal of Orthopsychiatry* 57, no. 1 (January 1987): 6–11. See also Daniel Goleman, "Deadlines for Change: Therapy in the Age of Reaganomics," *Psychology Today*, August 1981, 60–69.

91   On the role of the Palo Alto Group in founding and disseminating brief therapy (as well as the important contributions by other psychotherapists, such as Milton Erickson), see Coert F. Visser, "The Origin of the Solution-Focused Approach," *International Journal of Solution-Focused Practices* 1, no. 1 (2013): 10–17. On the economizing measures of managed health care as an engine for the growth of brief therapy, see Steven Stern, "Managed Care, Brief Therapy, and Therapeutic Integrity," *Psychotherapy: Theory, Research, Practice, Training* 30, no. 1 (1993): 162–75; and David A. Shapiro, Michael Barkham, William B. Stiles, Gillian E. Hardy, Anne Rees, Shirley Reynolds, and Mike Startup, "Time Is of the Essence: A Selective Review of the Fall and Rise of Brief Therapy Research," *Psychology and Psychotherapy: Theory, Research and Practice* 76, no. 3 (2003): 211–35. Shapiro and his coauthors privilege an alternate genealogy of brief therapy that extends from Franz Alexander and Thomas Morton French to David H. Malan and Peter Sifneos.

Chapter 3: Poeticizing Cybernetics

1   Claude Lévi-Strauss, "The Mathematics of Man," *International Bulletin of Social Sciences* 6, no. 4 (1954): 582.

2   Claude Lévi-Strauss, "The Mathematics of Man," 582.

3   On European humanists and cybernetics in the United States, see Gal-
    ison, "The Americanization of Unity." On Jakobson and the cybernetic
    turn in structural linguistics, see Lily Kay, *Who Wrote the Book of Life?:
    A History of the Genetic Code* (Stanford, CA: Stanford University Press,
    2000); Erhard Schüttpelz, "Quelle, Rauschen und Senke der Poesie:
    Roman Jakobsons Umschrift der Shannonschen Kommunikation," in
    Schnittstelle: Medien und Kommunikation, ed. Georg Stanitzek and Wil-
    jelm Voßkamp (Cologne: DuMont, 2001), 187–206; and Bernhard Sieg-
    ert, "Die Geburt der Literatur aus dem Rauschen der Kanäle: Zur Poetik
    der phatischen Funktion," in *Electric Laokoon: Zeichen und Medien, von
    der Lochkarte zur Grammatologie*, ed. Michael Franz, Bernhard Siegert,
    and Robert Stockhammer (Berlin: Akademie Verlag, 2007), 5–41.

4   On wartime cryptography, Shannon, and Wiener, see Mara Mills, "Media
    and Prosthesis: The Vocoder, the Artificial Larynx, and the History of
    Signal Processing," *Qui Parle* 21, no. 1 (Fall/Winter 2012): 112.

5   On the poetics of cybernetics, see also Lily Kay, "From Logical Neurons
    to Poetic Embodiments of Mind: Warren S. McCulloch's Project in Neu-
    roscience," *Science in Context* 14, no. 15 (2001): 591–614.

6   Edwards, *The Closed World*; Galison, "The Ontology of the Enemy."

7   American Telephone and Telegraph Company, "Our Many-Tongued
    Ancestors," *The Survey* 43, no. 19 (March 6, 1920): 715.

8   On the colonialist and theological backdrop to US communication infra-
    structures, see Nye, *American Technological Sublime*; and James W. Carey,
    "A Cultural Approach to Communications" (1989), in *Communication as
    Culture: Essays on Media and Society* (New York: Routledge, 2009), 14.

9   Ferdinand de Saussure, *Course in General Linguistics*, ed. Charles Bally
    and Albert Sechehaye, trans. Wade Baskin (New York: Philosophical Li-
    brary, 1959 [1916]), 22–23; Claude Lévi-Strauss, *Mythologiques*. Vols. 1–4.
    Paris: Plon, 1964–1971; Jacques Lacan, "A Theoretical Introduction to
    the Functions of Psychoanalysis in Criminology" (1950 speech, 1951 first
    published), in *Écrits: The First Complete Edition in English*, trans. Bruce
    Fink (New York: W. W. Norton, 2006), 102–22; Lacan, "Science and
    Truth," 739.

10 Roman Jakobson, *My Futurist Years*, ed. Bengt Jangfeldt, trans. Stephen
    Rudy (New York: Marsilio, 1997), 31, 278n63.

11 On Jakobson's refuge in the countryside, see remarks offered to the
    House Un-American Activities Committee, reported by Roman Jakobson,
    "Memoir" (recorded in 1977), in *My Futurist Years*, ed. Bengt Jangfeldt,
    trans. Stephen Rudy and Stephen Rudy (New York: Marsilio, 1997),
    281n78.

12 Jakobson, *My Futurist Years*, 83–85. See also Jindřich Toman, *The Magic
    of a Common Language: Jakobson, Mathesius, Trubetzkoy, and the Prague
    Linguistic Circle* (Cambridge, MA: MIT Press, 1995), 39, 87.

13   Jakobson, "Memoir," 65.

14   Roman Jakobson, "Principles of Historical Phonology" (1949 revision),
      in *On Language*, ed. Linda R. Waugh and Monique Monville-Burston
      (Cambridge, MA: Harvard University Press, 1990), 184–201.

15   Roman Jakobson, "The Generation That Squandered Its Poets (Excerpts)"
      (1930), trans. Dale E. Peterson, *Yale French Studies*, no. 39 (1967): 119. For
      a fuller and alternate translation, see Roman Jakobson, "On a Generation
      That Squandered its Poets" (1930), in *My Futurist Years*, ed. Bengt Jang-
      feldt, trans. Stephen Rudy and Stephen Rudy (New York: Marsilio, 1997),
      209–245.

16   Jakobson, "The Generation That Squandered Its Poets (Excerpts)," 125.

17   Jacques Derrida, "Force and Signification" (1963 lecture, 1967 print), in
      Writing and Difference, trans. Alan Bass (Chicago: University of Chicago
      Press, 1978), 5–6.

18   Jakobson, quoted in Stephen Rudy, "Introduction," in *My Futurist Years*,
      x; see also Toman, *The Magic of a Common Language*, 244.

19   Toman, *Magic of a Common Language*, 245.

20   On Eisenhower's probable intervention, see Jakobson, *My Futurist Years*,
      280–81n78. Reports of Jakobson's private testimony to the committee
      in April 1953 are recorded in a meeting he had with the FBI Boston Field
      Office: "Dr. Roman Jakobson," December 10, 1956, Federal Bureau of
      Investigation, Boston, Massachusetts, 105–706–27, p. 2 (apparently held
      in FBIA). Furnished to Geoghegan through a Freedom of Information Act
      request.

21   Tracy Kittredge, October 6, 1941, RF, RG 1.1, series 200, box 54, folder 63,
      RAC.

22   See correspondence in Rockefeller Foundation Collection, RG 2–1942,
      series 200, box 232, folder 1608, RAC.

23   Saussure, *Course in General Linguistics*, 41.

24   Saussure, *Course in General Linguistics*, 18.

25   Roman Jakobson, *Six Lectures on Sound and Meaning* (lectures ca. 1942)
      (Cambridge, MA: MIT Press, 1978), 16.

26   Jakobson, *Six Lectures*, 7–10, 17.

27   Jakobson also brought a Bell Labs engineer to the Libre to demonstrate
      their "visible speech" studies, which graphically visualized speech. See
      invitation cards and announcements for events held in 1944 and 1946 in
      Jakobson's file on the Linguistic Circle of New York, box 6, folder 74, RJP.
      On the visible speech research, see Mills, "Deaf Jam."

28   Roman Jakobson, "Notes on General Linguistics: Its Present State
      and Crucial Problems," RF, RG 1.2, series 200R, box 370, folder 3323
      (July 1949): 57–58, RAC.

29   On the intermingling of Slavic Studies and cybernetics, see Benjamin Pe-
      ters, "Toward a Genealogy of a Cold War Communication Sciences: The

Strange Loops of Leo and Norbert Wiener," *Russian Journal of Communication* 5, no. 1 (2013): 31–43, https://doi.org/10.1080/19409419.2013 .775544.

30 Roman Jakobson to Charles Fahs, February 22, 1950, box 6, folder 37, RJP.

31 *The Rockefeller Foundation: A Review for 1950 and 1951* (New York: Rockefeller Foundation, 1951), 77; US Department of State, "Foreign Affairs Research: Projects and Centers" (Washington, D.C., November 15, 1960), 120.

32 The Rockefeller Foundation, *The Rockefeller Foundation: A Review for 1950 and 1951* (New York: Rockefeller Foundation, 1951), 74.

33 Rockefeller, *Review for 1950 and 1951*, 12.

34 Roman Jakobson and Morris Halle, *Fundamentals of Language* (The Hague: Mouton, 1956), 17.

35 Jakobson and Halle, *Fundamentals of Language*, 3.

36 Roman Jakobson, "Results of a Joint Conference of Anthropologists and Linguists" (1952 address), in *Selected Writings II: Word and Language* (The Hague: Mouton, 1971), 556.

37 Jakobson, "Results of a Joint Conference," 559.

38 Beranek, Jakobson, and Locke to Technology Press, June 27, 1951, box 50, folder 29, RJP.

39 For the invitation to join the MIT faculty, see J. A. Stratton to Roman Jakobson, March 28, 1957, box 3, folder 67, RJP. For the invitation to join the *Information and Control* editorial board, see Jerome B. Wiesner to Roman Jakobson, May 29, 1956, box 50, folder 29, RJP. For the invitation to join the Center for Communication Sciences, see J. A. Stratton to Roman Jakobson, December 2, 1957, box 3, folder 63, RJP.

40 Stratton to Jakobson, March 28, 1957.

41 US Department of State, "Group Research Projects in Foreign Affairs and the Social Sciences" (Washington, DC: Bureau of Intelligence and Research, September 10, 1958), 89; Bauer, Inkeles, and Kluckhohn, *How the Soviet System Works*.

42 Roman Jakobson, "Closing Statements: Linguistics and Poetics" (1958), in *Style in Language* (Cambridge: MIT Press, 1960), 356.

43 Quoted in Slava Gerovitch, "Roman Jakobson und die Kybernetisierung der Linguistik in der Sowjetunion," in *Die Transformation des Humanen: Beiträge zur Kulturgeschichte der Kybernetik*, ed. Erich Hörl and Michael Hagner (Frankfurt: Suhrkamp, 2008), 243. I thank Professor Gerovitch for providing me with the original English text.

44 Jakobson, "Closing Statements," 356.

45 Jakobson, "Closing Statements," 358.

46 Jakobson, "Closing Statements," 370, 371.

47 Jakobson, "The Generation That Squandered Its Poets (Excerpts)," 125.

Chapter 4: Theory for Administrators

1 The catalog lists Dina Lévi-Strauss as an equal contributor, and her subsequent disappearance from Lévi-Strauss's accounts of the expedition has itself been a topic of growing interest in recent years. On Dreyfus in Brazil, see Ellen Spielmann, *Das Verschwinden Dina Lévi-Strauss' und der Transvestismus Mário de Andrades: Genealogische Rätsel in der Geschichte der Sozial- und Humanwissenschaften im modernen Brasilien* (Berlin: Wissenschaftlicher Verlag, 2003); and Luísa Valentini, "Um Laboratorio de Antropologia: O Encontro entre Mário de Andrade, Dina Dreyfus e Claude Lévi-Strauss, 1935–1938" (PhD diss., Universidade de São Paulo, 2013), https://www.teses.usp.br/teses/disponiveis/8/8134/tde-06062011-132611/pt-br.php.

2 Dina Dreyfus, "De la pensée prelogique à la pensée hyperlogique (à propos de la pensée sauvage)," *Mercure de France* (February 1963), 316. Translation of first, brief translation my own. Second translation taken from Emmanuelle Loyer, *Lévi-Strauss: A Biography*, trans. Ninon Vinsonneau and Jonathan Magidoff (Medford, MA: Polity, 2018), 373.

3 Claude Lévi-Strauss, "Social Structure," in *Anthropology Today: An Encyclopedic Inventory*, ed. A. L. Kroeber (Chicago: University of Chicago Press, 1953), 538.

4 There are important distinctions in French between *ethnography*, *ethnology*, and *anthropology*, which vary according to time and place, but which are less marked in contemporary English. Fournier summarizes them well: "Anthropology is the comparative study of beliefs and institutions, understood as the foundation of social structures. Ethnography entails the description of ethnic groups, while ethnology studies these same groups in terms of the unity of their linguistic, economic, and social structure and in terms of their evolution. Mauss played somewhat on these distinctions, embracing both sociology and anthropology or ethnology." Marcel Fournier, *Marcel Mauss: A Biography*, trans. Jane Marie Todd (Princeton, NJ: Princeton University Press, 2006 [1994]), 234.

5 For more on technocracy and technical ideals in France, see Theodore M. Porter, *Trust in Numbers: The Pursuit of Objectivity in Science and Public Life* (Princeton, NJ: Princeton University Press, 1996); Ken Alder, *Engineering the Revolution: Arms and Enlightenment in France, 1763–1815* (Princeton, NJ: Princeton University Press, 1997); Bruno Belhoste, *La formation d'une technocratie: L'école polytechnique et ses élèves de la révolution au second empire* (Paris: Belin, 2003).

6 For a more recent example of French thought valorizing technical power while contesting technical neutrality, see Bruno Latour, *Reassembling the*

Social: An Introduction to Actor-Network-Theory (Oxford: Oxford University Press, 2005).

7   Robert Brain, "Standards and Semiotics," in *Inscribing Science: Scientific Texts and the Materiality of Communication*, ed. Timothy Lenoir (Stanford, CA: Stanford University Press, 1998), 251.

8   Fournier, *Marcel Mauss*, 166.

9   Fournier, *Marcel Mauss*, 237.

10   For more on French notions of the collective, see Larry Siedentrop, "Two Liberal Traditions," in *The Idea of Freedom: Essays in Honour of Isaiah Berlin*, ed. Alan Ryan (Oxford: Oxford University Press, 1979), 153–74.

11   Fournier, *Marcel Mauss*, 246.

12   Fournier, *Marcel Mauss*, 247–48.

13   Fournier, *Marcel Mauss*, 248.

14   Marcel Mauss, "Les sciences sociales à Paris vues par Marcel Mauss," *Revue française de sociologie* 26, no. 2 (1985 [1929]): 343–51.

15   Fournier, *Marcel Mauss*, 256.

16   Fournier, *Marcel Mauss*, 256; Brigitte Mazon, "La fondation Rockefeller et les sciences sociales en France, 1925–1940," *Revue française de sociologie* 26, no. 2 (1985): 323–27, https://doi.org/10.2307/3321579.

17   Fournier, *Marcel Mauss*, 293. Among those who benefited from this largesse was Lévi-Strauss, at that time a young ethnographer whose missions in the Brazilian jungle (later recounted in *Tristes Tropiques*) were underwritten by the Rockefeller-funded Institut d'ethnologie. The general paucity of fieldwork funding meant these investigations were the first, and ultimately the only, ethnographic fieldwork Lévi-Strauss had the opportunity to carry out early in his career. In fulfillment of the Rockefeller strategy, however, those individualistic and concrete observations became a cornerstone of Lévi-Strauss's subsequent social, political, and ethical reflections. Brigitte Mazon, *Aux origines de l'école des hautes études en sciences sociales: Le rôle du mécénat américain (1920–1960)* (Paris: Les éditions du Cerf, 1988), 57.

18   T. B. Kittredge, "General policy affecting the future of social sciences programs in Europe," January 23, 1935, RF, "Program and Policy 1933–1936," RG 3.1, series 910, box 1, folder 3, RAC.

19   Marcel Mauss, *The Gift: The Form and Reason for Exchange in Archaic Societies*, trans. W. D. Halls (New York: Routledge, 1990).

20   On the political context of Mauss's essay, see Simon Torracinta, "The Peripeteia of *The Gift*," review of *Gift Exchange: The Transnational History of a Political Idea*, by Grégoire Mallard, *History of Anthropology Review* 44 (October 20, 2020), https://histanthro.org/reviews/peripeteia-of-the-gift/.

21   Mauss, *The Gift*, 98. Translation modified.

22   See Mary Douglas, "Preface," in Mauss, *The Gift*, xiii–xv. For overviews of the contrasts between Anglo and French liberalism, see Siedentrop, "Two Liberal Traditions."

23   On the ELHE, see Aristide R. Zolberg and Agnès Callamard, "The École Libre at the New School, 1941–1946," *Social Research* 65, no. 4 (Winter 1998): 921–51; and François Chaubet and Emmanuelle Loyer, "L'École libre des hautes études de New York: Exil et résistance intellectuelle (1942–1946)," *Revue Historique* 302, no. 4/616 (October–December 2000): 939–72. On Lévi-Strauss's time in New York, see Jeffrey Mehlman, *Emigré New York: French Intellectuals in Wartime Manhattan, 1940–1944* (Baltimore, MD: Johns Hopkins University Press, 2000). For the ELHE in the context of francophone exilic New York, see Emmanuelle Loyer, *Paris à New York: Intellectuels et artistes français en exil (1940–1947)* (Paris: Bernard Grasset, 2005).

24   Tracy Kittredge, October 6, 1941, RF, RG 1.1, series 200, box 54, folder 63, RAC.

25   See correspondence in RF, RG 2-1942, series 200, box 232, folder 1608, RAC.

26   For records of these and other seminars, see RF, RG 1.1, series 200, box 54, folder 634, RAC.

27   RF, RG 1.1, series 200, box 54, folder 634, RAC; and "Statement on the Commission for the Study of the Reforms of the French State," untitled folder, box 3, École Libre Papers, New School for Social Research Archives, New York (hereafter ELP).

28   "Statement on the Commission for the Study of the Reforms of the French State."

29   Claude Lévi-Strauss, "Le socialisme et la colonialisation," *L'étudiant socialiste*, October 1929, 7–8; Vincent Chambarlhac, "Lévi-Strauss en socialisme," *Cahiers d'histoire: Revue d'histoire critique*, no. 101 (2007): 83–99; Alexandre Pajon, *Claude Lévi-Strauss: Politique* (Descartes: Éditions Faustroll, 2007).

30   Lévi-Strauss, *Tristes Tropiques*, (New York: Atheneum, 1975 [1955]): 25–26.

31   Lévi-Strauss, *Tristes Tropiques*, 35.

32   Lévi-Strauss, *Tristes Tropiques*, 35.

33   Lévi-Strauss, *Tristes Tropiques*, 36.

34   Hoover to Special Agent in Charge, April 17, 1941, FBIA. Special thanks to John Cook, formerly of Gawker.com, who furnished me with copies of Lévi-Strauss's FBI files, which he secured through a Freedom of Information Act request.

35   Hoover to SAO (New York), March 3, 1942, FBIA.

36   Loyer, *Lévi-Strauss*, 195–96, 216.

37   Claude Lévi-Strauss and Didier Eribon, *De près et de loin* (Paris: Odile Jacob, 1988), 61; "Claude Levi-Strauss," Collection Refugee Scholars—New

School, RG 1.1, Series 200, Box 54, Folder 632, RAC; "Claude Lévi-Strauss" (folder), ELP. On Latin American influence as a strategic rationale for establishing the École libre, see also Alvin Johnson to Henri Grégoire, September 30, 1941, "Ecole Libre Papers / Archives: Alvin Johnson & Zambelli Project, Alexandre Koyré," Box 3, ELP.

38   See Mehlman, *Émigré New York*, 181–82; and Stephen Rudy, "Jakobson et Lévi-Strauss à New York (1941–1945), and Then Those Infamous Cats," in *Claude Lévi-Strauss*, ed. Michel Izard (Paris: Herne, 2004), 120–24.

39   Mehlman, *Émigré New York*, 181–2.

40   Patrick Wilcken, *Claude Lévi-Strauss: The Poet in the Laboratory* (New York: Penguin, 2010), 154.

41   Lévi-Strauss to John Marshall, August 1944, RF, RG 1.1, series 200, box 52, folder 610, RAC.

42   Pierre Bourdieu, "Préface," in *Aux origines de l'école des hautes études en sciences sociales: Le rôle du mécénat américain (1920–1960)* (Paris: Les éditions du Cerf, 1988), i.

43   Quoted in Loyer, *Lévi-Strauss*, 280.

44   Jakobson to Lévi-Strauss, January 16, 1949, in Roman Jakobson and Claude Lévi-Strauss, *Correspondance 1942–1982*, ed. Emmanuelle Loyer and Patrice Maniglier (Paris: Éditions du Seuil, 2018), 102.

45   See also Jakobson to Lévi-Strauss, January 26, 1949, and February 14, 1950, in Jakobson and Lévi-Strauss, *Correspondance,* 102, 127.

46   Lévi-Strauss to Jakobson, February 7, 1949, in Jakobson and Lévi-Strauss, *Correspondance,* 103.

47   Charles Fahs, entry for September 8, 1949, RF, diary, RG-2 1949, series 500R, box 483, folder 3104, RAC. Lévi-Strauss occasionally changed the spelling of his name, which may account for Fahs's spelling of the name as "Levy-Strauss."

48   Claude Lévi-Strauss, "Language and the Analysis of Social Laws" (1951), in *Structural Anthropology* (New York: Basic Books, 1963), 57–58.

49   As MIT linguist Noam Chomsky later recalled, in 1954 it did not seem at all unreasonable when an anthropological linguist told him he would postpone working through his vast collection of materials "because within a few years it would surely be possible to program a computer to construct a grammar from a large corpus of data by the use of techniques that were already fairly well formalized." It is possible this was an allusion to Lévi-Strauss. Noam Chomsky, *Language and Mind* (Cambridge: Cambridge University Press, 2006 [1968]), 2.

50   Claude Lévi-Strauss, *Introduction to the Work of Marcel Mauss*, trans. Felicity Baker (London: Routledge, 1987 [1950]), 36.

51   Lévi-Strauss, *Introduction*, 56.

52   Lévi-Strauss, *Introduction*, 44.

53   Lévi-Strauss, *Introduction*, 44.

54   Lévi-Strauss, *Introduction*, 8; Marcel Mauss, "Techniques of the Body," *Economy and Society* 2, no. 1 (February 1973 [1934]): 70–88.

55   Lévi-Strauss, *Introduction*, 8.

56   Theodora Vardouli, "Dissimilar at First Sight: Structural Abstraction and the Promises of Isomorphism in 1960s Architectural Theory," in *Instabilities and Potentialities: Notes on the Nature of Knowledge in Digital Architecture*, ed. Chandler Ahrens and Aaron Sprecher (New York: Routledge, 2019), 209–20.

57   On complementarities in structural reasoning across the social sciences and anthropology, see Theodora Vardouli, "Dissimilar at First Sight," 209–20.

58   Loyer, *Lévi-Strauss*, 308.

59   Loyer, *Lévi-Strauss*, 307.

60   John Marshall, notes from interview with Lévi-Strauss, September 30, 1949 (Marshall's paraphrase rather than a direct quote), RF, RG 2-1949, series 500R, box 483, folder 3104, RAC.

61   Lévi-Strauss, "The Mathematics of Man," 583, 587–86.

62   Wiktor Stoczkowski, "Claude Lévi-Strauss and UNESCO," *Courrier de l'UNESCO* 5 (2008): 5–7; Wiktor Stoczkowski, "The 'Unconceivable Humankind' to Come: A Portrait of Lévi-Strauss as a Demographer," *Diogenes* 60, no. 2 (2013): 79–92, https://doi.org/10.1177/0392192114568268.

63   Claude Lévi-Strauss, "Louis-Ferdinand Céline: Voyage au bout de la nuit" (1933), in *Claude Lévi-Strauss*, ed. Michel Izard (Paris: Herne, 2004), 24; my translation.

64   Lévi-Strauss, *Tristes Tropiques*, 284.

65   Claude Lévi-Strauss, review of "La grande épreuve des démocraties: Essai sur les principes démocratiques—leur nature—leur histoire—leur valeur philosophique," by Julien Benda, *Renaissance* 2–3 (1944): 324, my translation.

66   Lévi-Strauss, *Tristes Tropiques*, 397. Translation modified slightly, to restore the non-capitalization of entropology (*entropologie*) in the French original.

67   Claude Lévi-Strauss, "The Place of Anthropology in the Social Sciences" (1954), in *Structural Anthropology* (New York: Basic Books, 1976), 366.

68   Lévi-Strauss, "Place of Anthropology," 367.

69   On game theory in *Race and History*, and Lévi-Strauss's appeals to information theory and cybernetics more generally, see Christopher Johnson, *Claude Lévi-Strauss: The Formative Years* (New York: Cambridge University Press, 2003). On the reception of game theory by Lévi-Strauss and his contemporaries in 1950s France, see Rabia Nessah, Tarik Tazdaït, and Merhdad Vahabi. "The Game Is Afoot: The French Reaction to Game Theory in the 1950s," *History of Political Economy* 53, no. 2 (2021): 243–78.

70   Claude Lévi-Strauss, *Race and History* (Paris: UNESCO, 1952), 48.

71   Claude Lévi-Strauss, "Diogène Couché" (1955), *Cités* 81, no. 1 (2020): 138; my translation.

72   For more on the Ford Foundation in Europe, including its efforts to modernize anthropology, see Gemelli, *The Ford Foundation and Europe* (Brussels: European Interuniversity Press, 1998); and Francis X. Sutton, "The Ford Foundation's Transatlantic Role and Purposes, 1951–81," *Review (Fernand Braudel Center)* 24, no. 1 (2001): 77–104.

73   In light of some changes from the original publication to its subsequent edition in *Structural Anthropology*, I cite the former. Claude Lévi-Strauss, "The Structural Study of Myth," *Journal of American Folklore* 68, no. 270 (1955): 439, https://doi.org/10.2307/536768.

74   Lévi-Strauss, "Structural Study of Myth," 443.

75   Lévi-Strauss, "Structural Study of Myth," 443.

76   Lévi-Strauss, "Structural Study of Myth," 443.

77   Lévi-Strauss, "Structural Study of Myth," 443.

78   E. F. D'Arms diary entry, February 17, 1950, RF, RG 2-1950, series 500R, box 500, folder 3347, RAC. The original identifying record was partially illegible. I have constructed this citation with the generous assistance of the RAC archivists and am reasonably confident it is correct. Thanks, in particular, to Bethany Antos for help finding the correct citation.

79   Schützenberger to Wiener, November 10, 1951, folder 143, NWP.

80   For more on CENIS's activities related to communications and the CIA, see Light, *From Warfare to Welfare*, 166; Alan A. Needell, "Project Troy and the Cold War Annexation of the Social Sciences," in *Universities and Empire: Money and Politics in the Social Sciences during the Cold War*, ed. Christopher Simpson (New York: New Press, 1998), 3–38. On foundations, CENIS, and US Cold War policy, see also James Schwoch, *Global TV: New Media and the Cold War, 1946–69* (Urbana: University of Illinois Press, 2009), 63–65.

81   Claude Lévi-Strauss, "Comparative Religions of Nonliterature Peoples" (1968), in *Structural Anthropology, Volume II* (New York: Basic Books, 1976), 62.

82   Alexander Campolo, "Signs and Sight: Jacques Bertin and the Visual Language of Structuralism," *Grey Room*, no. 78 (Winter 2020): 34–65, https://doi.org/10.1162/grey_a_00286.

83   Loyer, *Lévi-Strauss*, 385.

84   Clellan S. Ford, "Human Relations Area Files: 1959–1969. A Twenty-Year Report" (New Haven, CT: Human Relations Area Files, 1970), 7.

85   Ford, "Human Relations Area Files," 27.

86   Office of the Chief of Naval Operations, "Military Government Handbook: Kurile Islands" (Washington, DC: Office of the Chief of Naval Operations, November 1, 1943), iii, http://archive.org/details/01150190R.nlm.nih.gov.

87   Office of the Chief of Naval Operations, "Military Government Handbook," viii and 14.

88   Claude Lévi-Strauss, "Postscript to Chapter XV" (1958), in *Structural Anthropology* (New York: Basic Books, 1976), 343.

89   Lévi-Strauss, "Postscript to Chapter XV," 343.

Chapter 5: Learning to Code

Epigraph: Jacques Derrida, "Implications: Interview with Henry Ronse" (1967), in *Positions*, trans. Alan Bass (Chicago: University of Chicago Press, 1981), 5. Translation slightly modified.

1   Nathalie Sarraute, *Portrait of a Man Unknown*, trans. Maria Jolas (New York: George Braziller, 1958 [1948]). For a contemporary American response, see Gerald Sykes, "Drab, Bored and Brutish," *New York Times*, August 10, 1958.

2   Alain Robbe-Grillet, "On Several Obsolete Notions" (1957), in *For a New Novel: Essays on Fiction*, trans. Richard Howard (Evanston, IL: Northwestern University Press, 1989), 28–29.

3   Jacques Lacan, "Jeunesse de Gide ou la lettre et le désir" (1958), in *Écrits* (Paris: Éditions du Seuil, 1966), 739. Alternate English rendering in Jacques Lacan, "The Youth of Gide, or the Letter and Desire," in *Écrits: The First Complete Edition in English*, trans. Bruce Fink (New York: W. W. Norton, 2006), 623.

4   Bell, "Post-Industrial Society," 28.

5   Exceptional works on (post)structuralism and cybernetics include Hayles, *Chaos Bound*, esp. 175–208; Lafontaine, *L'Empire cybernétique*; Cusset, "Cybernétique et « théorie française »;" Liu, *Freudian Robot*; Sack, "Une Machine à raconter"; and Segal's *Le zéro et le un (vol. 1)*, 485–532.

6   François Cusset, *French Theory: How Foucault, Derrida, Deleuze, and Co. Transformed the Intellectual Life of the United States*, trans. Jeff Fort, Josephine Berganza, and Marlon Jones (Minneapolis: University of Minnesota Press, 2008). Exceptions, such as Stuart Hall's wry readings of Althusser and Barthes, also exist, though the extent to which his negotiated reading of "encoding/decoding" remained negotiated in its later popularization is up for debate.

7   Paul Ricoeur, *Freud and Philosophy: An Essay on Interpretation* (New Haven, CT: Yale University Press, 1970); Vincent Descombes, *Modern French Philosophy* (Cambridge: Cambridge University Press, 1980).

8   Paul de Man, "Roland Barthes and the Limits of Structuralism" (ca. 1972), *Yale French Studies*, no. 77 (1990): 181, https://doi.org/10.2307/2930153.

9 Quoted in Norbert Wiener, *The Human Use of Human Beings: Cybernetics and Society* (Boston: Da Capo Press, 1988 [1950]), 179; original in Pére Dominique Dubarle, "Vers la machine à gouverner," *Le Monde*, December 28, 1948, 47–49.

10 Quoted in Wiener, *Human Use of Human Beings*, 180.

11 Pierre Bourdieu, *The State Nobility: Elite Schools in the Field of Power*, trans. Lauretta C. Clough (Cambridge: Polity, 1996 [1989]), 336–37.

12 Bourdieu, *The State Nobility*, 337.

13 Richard Kuisel, *Seducing the French: The Dilemma of Americanization* (Berkeley: University of California Press, 1993), 70, 85.

14 On nuclear power and the radiance of postwar France, see Gabrielle Hecht, *The Radiance of France: Nuclear Power and National Identity after World War II* (Cambridge, MA: MIT Press, 2009); on the context of postwar social science, especially structuralism, see Kristin Ross, *Fast Cars, Clean Bodies: Decolonization and the Reordering of French Culture* (Cambridge, MA: MIT Press, 1995), 157–96; on management and human engineering, see Luc Boltanski, "Visions of American Management in Post-war France" (1981), trans. Alexandra Russell, *Theory and Society* 12, no. 3 (1983): 375–403; on postwar reforms in France, see Kuisel, *Seducing the French*.

15 Lucile Dumont, "Literary Theorists in and beyond French Academic Space (1960–1970s)," *Sociological Review* 68, no. 5 (2020): 1113, https://doi.org/10.1177/0038026120916119; Lucile Dumont, "From Sociology to Literary Theory: The Disciplinary Affiliations of Literature in Section VI of the École Pratique des Hautes Études (EPHE), 1956–1975," trans. Jean-Yves Bart, *Symbolic Goods* 3 (2018), https://www.biens-symboliques.net/289?lang=en.

16 The original names of the centers were the Laboratoire de cartographie, Groupe de mathématique sociale et de statistique, Centre d'études des communications de masse, Centre d'études transdisciplinaires, and Centre de sociologie européenne.

17 Ross, *Fast Cars, Clean Bodies*, 187.

18 Ross, *Fast Cars, Clean Bodies*, 186.

19 François Dosse, *History of Structuralism: The Rising Sign, 1945–1966*, trans. Deborah Glassman (Minneapolis: University of Minnesota Press, 1998 [1991]), 62.

20 See Johannes Angermuller, *Why There Is No Poststructuralism in France: The Making of an Intellectual Generation* (London: Bloomsbury Academic, 2015), esp. 28–30, 45–55.

21 Niilo Kauppi, *The Making of an Avant-Garde: Tel Quel* (Berlin: Mouton de Gruyter, 1994), 108–9.

22 Regarding Ford Foundation funding, see Ronan Le Roux, *Une histoire de la cybernétique en France (1948–1975)* (Paris: Classiques Garnier, 2018),

592–93; regarding CENIS (MIT) funding, see Jakobson and Lévi-Strauss, *Correspondance 1942–1982*, 173.

23   See Lévi-Strauss, letter to Millikan, January 7, 1953, box 50, folder 29, RJP; and Le Roux, *Une histoire*, 620–21. I am assuming "A. Ross" of the University of Birmingham referred, in fact, to W. Ross Ashby.

24   Lévi-Strauss, letter to Jakobson, May 5, box 12, folder 45, RJP. There was no year present on the letter, but the contents suggest it was written in 1953; my translation.

25   Jacques Riguet, "Une analyse indolore: Propos recueillis par Deborah Gutermann-Jacquet," *Le diable probablement* 9 (2011): 98–102.

26   Dennis Duncan, *The Oulipo and Modern Thought* (Oxford: Oxford University Press, 2019), 32. For a masterful account of how informa-tion theory and related notions inflected the work of one member, Georges Perec, see Marc Kohlbry, "Digital Index: Control Poetics in *Die Maschine*," *Symploke* 28, no. 1 (November 24, 2020): 83–99; and Marc Kohlbry, "Feedback, Information, and the Fabrication of Spatialisme," *MLN* 135, no. 4 (2020): 859–87, https://doi.org/10.1353/mln.2020.0056.

27   Loyer, *Lévi-Strauss*, 406.

28   Lévi-Strauss and Eribon, *De près et de loin*, 24; my translation.

29   Quoted in Dosse, *History of Structuralism*, 9.

30   Michel Foucault, "Truth and Power" (1977), in *Power/Knowledge: Selected Interviews and Other Writings*, ed. Colin Gordon (New York: Pantheon, 1980), 128, 131.

31   Sunil Khilnani, *Arguing Revolution: The Intellectual Left in Postwar France* (New Haven, CT: Yale University Press, 1993), 83–84.

32   Niilo Kauppi, *French Intellectual Nobility: Institutional and Symbolic Transformations in the Post-Sartrian Era* (Albany: SUNY Press, 1996), 119–20; see also the list of Barthes's seminars in Lucy O'Meara, ed., "List of Roland Barthes's Seminars and Lecture Courses at the École Pratique des Hautes Études and the Collège de France, 1963–1980," in *Roland Barthes at the Collège de France* (Liverpool: Liverpool University Press, 2012), 205–6, https://www.cambridge.org/core/books/roland-barthes -at-the-college-de-france/list-of-roland-barthess-seminars-and-lecture -courses-at-the-ecole-pratique-des-hautes-etudes-and-the-college-de -france-19631980/E41854AD9F9D503A619932AED1416E9D.

33   Kauppi, *French Intellectual Nobility*, 77.

34   Jacques Lacan, "The Mirror Stage as Formative of the I Function" (1949 version), in *Écrits: The First Complete Edition in English*, trans. Bruce Fink (New York: W. W. Norton, 2006), 75–81. For more on the interwar conception of the body as broken, traumatized, and shocked, see Walter Benjamin, "The Work of Art in the Age of Its Technological Reproduc-ibility" (1935 version), trans. Michael W. Jennings, *Grey Room*, no. 39 (Spring 2010): 11–37; and Susan Buck-Morss, "Aesthetics and Anaesthetics:

Walter Benjamin's Artwork Essay Reconsidered," *October*, no. 62 (1992): 3–41, https://doi.org/10.2307/778700.

35   Lacan, "Science and Truth," 730. Lacan is referencing Georges Canguilhem's joke, in another text, that "When one leaves the Sorbonne by the street Saint-Jacques, one can ascend or descend; if one ascends, one approaches the Pantheon, the conservatory of great men; but if one descends, one heads directly to the Police Department" See in "What Is Psychology?," trans. David M. Peña-Guzmán, *Foucault Studies*, no. 21 (June 2016): 213.

36   See General Correspondence Folder, RF, RG 2-1942, series 200, box 232, folder 1608, RAC.

37   Elisabeth Roudinesco, *Jacques Lacan*, trans. Barbara Bray (New York: Columbia University Press, 1997 [1993]), 260; Louis Althusser, *Psychoanalysis and the Human Sciences* (lectures in 1963–64), trans. Steven Rendall (New York: Columbia University Press, 2016), 40.

38   Heyck, *Age of System*, 1–17.

39   Althusser, *Psychoanalysis*, 41. Lacan himself once said of his auditors, "What do they come for? What do they think? Their eyes are so blank. And all those tape recorders are like guns trained on me! They don't understand; I'm absolutely convinced they don't understand anything. They come just so as to be able to say they're writing a book on Lacan, a biography of Lacan." Roudinesco, *Jacques Lacan*, 353.

40   Quoted in Roudinesco, *Jacques Lacan*, 211; original in Lévi-Strauss and Eribon, *De près et de loin*, 107–8.

41   Quoted in Dany Nobus, *Critique of Psychoanalytic Reason: Studies in Lacanian Theory and Practice* (New York: Routledge, 2022), 12.

42   Excerpts from an Italian interview quoted in Roudinesco, *Jacques Lacan*, 332.

43   For a fuller overview of the text, especially the revised and canonical form that appeared in *Écrits*, see John P. Muller and William J. Richardson, "Lacan's Seminar on 'The Purloined Letter': Overview," in *The Purloined Poe: Lacan, Derrida and Psychoanalytic Reading* (Baltimore, MD: Johns Hopkins University Press, 1988), 55–76. On aspects of the text's computational elements, see John Johnston, *The Allure of Machinic Life: Cybernetics, Artificial Life, and the New AI* (Cambridge, MA: MIT Press, 2008), 75–103; S. Berlin Brahnam, "On Lacan's Neglected Computational Model and the Oedipal Structure," *S: Journal of the Circle for Lacanian Ideology Critique*, nos. 10–11 (2017–18): 261–74; and S. Berlin Brahnam, "Primordia of Après-Coup, Fractal Memory, and Hidden Letters," *S: Journal of the Circle for Lacanian Ideology Critique*, nos. 10–11 (2017–18): 202–44.

44   For more on the stakes of cryptography and language in this period, see Dileonardi, "Cryptographic Reading."

45 D. W. Hagelbarger, "SEER, A SEquence Extraction Robot," *IRE Trans-actions on Electronic Computers* EC-5, no. 1 (March 1956): 1–4; Claude E. Shannon, "A Mind-Reading (?) Machine" (1953 memo), in *Claude El-wood Shannon: Collected Papers*, ed. N. J. A. Sloane and Aaron D. Wyner (Piscataway, NJ: IEEE Press, 1993), 688–90.

46 Jacques Lacan, *The Ego in Freud's Theory and in the Technique of Psycho-analysis, 1954–1955*, trans. Sylvan Tomaselli (New York: W. W. Norton, 2006), 190.

47 Lacan, *The Ego*, 47. Translation modified.

48 Jacques Lacan, "The Freudian Thing or the Meaning of the Return to Freud in Psychoanalysis" (1955 lecture, 1956 first published), in *Écrits: The First Complete Edition in English*, trans. Bruce Fink (New York: W. W. Norton, 2006), 344.

49 Jacques Lacan, "The Function and Field of Speech and Language in Psychoanalysis" (1953 speech, 1956 first published), in *Écrits: The First Complete Edition in English*, trans. Bruce Fink (New York: W. W. Norton, 2006), 247.

50 Joan Copjec, *Read My Desire: Lacan against the Historicists* (Cambridge, MA: MIT Press, 1994), 18.

51 Axel L. Wenner-Gren, "Address of Welcome," in *An Appraisal of Anthro-pology Today*, ed. Sol Tax, Loren C. Eiseley, Irving Rouse, and Carl F. Voegelin (Chicago: University of Chicago Press, 1953), xiii–xiv.

52 Lévi-Strauss, "Social Structure," 528, 538.

53 Mead, in "Problems of Process: Results" (roundtable discussion), in *An Appraisal of Anthropology Today*, ed. Sol Tax, Loren C. Eiseley, Irving Rouse, and Carl F. Voegelin (Chicago: University of Chicago Press, 1953), 111.

54 Northrop, in "Pattern in Biology, Linguistics, and Culture," in *An Ap-praisal of Anthropology Today*, ed. Sol Tax Loren C. Eiseley, Irving Rouse, and Carl F. Voegelin, (Chicago: University of Chicago Press, 1953), 315.

55 Lévi-Strauss, in "Pattern in Biology, Linguistics, and Culture," 321.

56 Georges Gurvitch, "Le concept de structure sociale," *Cahiers internation-aux de sociologie* 19 (July–December 1955): 3–44.

57 Alain Touraine, "Le traitement de la société globale dans la sociologie américaine contemporaine," *Cahiers internationaux de sociologie* 16 (June 1954): 126–45; Henri Lefebvre, "La notion de totalité dans les sciences sociales," *Cahiers internationaux de sociologie* 18 (January 1955): 55–77.

58 Henri Lefebvre, "Marxisme et la théorie de l'information" (1958), in *Au-delà du structuralisme* (Paris: Anthropos, 1971), 72.

59 Alain Touraine, "Entreprise et bureaucratie," *Sociologie du travail* 1, no. 1 (1959): 58–71, https://doi.org/10.3406/sotra.1959.1003; Jean Meynaud, "Qu'est-ce que la technocratie?," *Revue économique* 11, no. 4

(1960): 497–526, https://doi.org/10.3406/reco.1960.407420; Jean Meynaud, *Technocratie et politique* (Lausanne: Études de science politique, 1960); Nora Mitrani, "Bureaucratie et technicité," *Arguments* 4, no. 17 (1960): 28–30; Pierre Fougeyrollas, "Bureaucratie et techno-cratie," *Arguments* 4, no. 17 (1960): 31–32; Claude Lefort, "Qu'est-ce que la bureaucratie?," *Arguments* 4, no. 17 (1960): 64–81; Alain Touraine, "L'aliénation bureaucratique," *Arguments* 4, no. 17 (1960): 21–27; Hecht, *The Radiance of France.*

60   Quoted in Christine Quinan, "Technocrats and Tortured Bodies: Simone de Beauvoir's *Les Belles Images*," *Women: A Cultural Review* 25, no. 3 (July 3, 2014): 258, https://doi.org/10.1080/09574042.2014.981380. Translation slightly modified.

61   Jean-Paul Sartre, "Jean-Paul Sartre Répond," *L'arc* 30 (October 1966), 94, 95; my translation.

62   Pierre Rottenberg, "Lectures de codes," in *Théorie d'ensemble* (Paris: Édi-tions du Seuil, 1968), 179, 182.

63   Raymond Queneau, "Présentation de l'encyclopédie" (1955), in *Bords: Mathématiciens précurseurs encyclopédistes* (Paris: Hermann, 1963), 85; my translations here and below.

64   Queneau, "Présentation de l'encyclopédie," 102; my translation.

65   Khilnani, *Arguing Revolution*, 84.

66   Quinan, "Technocrats and Tortured Bodies."

67   Simone de Beauvoir, *Les belles images*, trans. Patrick O'Brian (London: Fontana/Collins, 1975 [1966]), 36.

68   De Beauvoir, *Les belles images*, 61.

69   De Beauvoir, *Les belles images*, 79.

70   Lévi-Strauss, *The Savage Mind*, 268.

71   Clifford Geertz, "The Cerebral Savage: On the Work of Claude Lévi-Strauss" (1967), in *The Interpretation of Cultures: Selected Essays* (New York: Basic Books, 1973), 359. See also Poornima Paidipaty, "'Tortoises All the Way Down': Geertz, Cybernetics and 'Culture' at the End of the Cold War," *Anthropological Theory* 20, no. 1 (2020): 97–129.

72   Edward Baring, "Liberalism and the Algerian War: The Case of Jacques Derrida," *Critical Inquiry* 36, no. 2 (Winter 2010): 239–61, https://doi.org/10.1086/648525.

73   Dina Dreyfus, "De la pensée prelogique," *Mercure de France* (February 1963), 316; my translation here and below.

74   Dreyfus, "De la pensée prelogique," 310.

75   Lévi-Strauss, *The Savage Mind*, 247.

76   Michel Foucault and Alain Badiou, "Philosophie et psychologie" (1965), in *Dits et écrits, tome 1* (Paris: Editions Gallimard, 1994), 447; my translation.

77   Michel Foucault, "Linguistique et sciences sociales" (1969), in *Dits et écrits, tome 1* (Paris: Editions Gallimard, 1994), 825.

78   Foucault, "Linguistique et sciences sociales," 825.

79   Dina Dreyfus, Alain Badiou, Jean Hyppolite, Georges Mounin, Pierre
      Bourdieu, and Jean Laplanche, *Emissions de philosophie 1966–1967
      consacrées au langage*, Paris: Institut pédagogique, Dossiers pédagogiques
      de la radiotélevision scolaire, 1967.

80   Tamara Chaplin Matheson, "Embodying the Mind, Producing the Na-
      tion: Philosophy on French Television," *Journal of the History of Ideas* 67,
      no. 2 (2006): 315–42.

81   For a critical discussion of these same terms by Foucault, see Michel Fou-
      cault, "Message ou bruit" (1966), in *Dits et écrits, tome 1* (Paris: Editions
      Gallimard, 1994), 557–60.

82   Lacan, *The Ego*, 83.

83   Some French scholars such as Georges Mounin and A. J. Greimas contin-
      ued to invoke code in neutral terms. They also tended to stress the limits
      of the informatic concept of code, applying it a more more circumscribed
      manner.

84   E. Colin Cherry, Morris Halle, and Roman Jakobson, "Toward the
      Logical Description of Languages in Their Phonemic Aspect," *Language*
      29, no. 1 (March 1953): 34–46; Jakobson and Halle, *Fundamentals of
      Language*; Gunnar Fant, "Phonetics and Phonology in the Last 50 Years,"
      paper presented at From Sound to Sense: 50+ Years of Discoveries in
      Speech Communication, MIT, Research Laboratory of Electronics,
      June 11–13, 2004, http://www.rle.mit.edu/soundtosense/conference
      /pdfs/invitedspeakers/fant%20paper.pdf.

85   *Communications*, "Vie des centres 1960–1961," *Communications* 1, no. 1
      (1961): 226–30; Tiphaine Samoyault, *Barthes: A Biography*, trans. Andrew
      Brown (Cambridge: Polity, 2017 (2015)), 203.

86   Samoyault, *Barthes*, 203.

87   Samoyault, *Barthes*, 203.

88   See, for example, Georges Canguilhem, "Machine and Organism" (1947
      lecture), trans. Mark Cohen and Randall Cherry, in *Incorporations*, ed.
      Jonathan Crary and Sanford Kwinter (New York: Zone, 1992), 45–69;
      Stefanos Geroulanos, "Violence and Mechanism: Georges Canguilhem's
      Overturning of the Cartesian Legacy," *Qui Parle* 24, no. 1 (Fall/Winter
      2015): 141, https://doi.org/10.5250/quiparle.24.1.0125; Jacques Lacan,
      "Discours de Rome" (1953), in *Autres écrits* (Paris: Éditions du Seuil,
      2001), 148.

89   Roland Barthes, *Elements of Semiology*, trans. Annette Lavers and Colin
      Smith (New York: Hill and Wang, 1968), 9.

90   Barthes, *Elements of Semiology*, 9.

91   Roland Barthes, *Album: Unpublished Correspondence and Texts*, ed. Éric
      Marty, trans. Jody Gladding (New York: Columbia University Press,
      2018), xxii–xxiii.

92 Barthes, *Album*, 179.

93 Barthes, *Album*, 107.

94 Barthes, *Album*, 113.

95 Barthes, *Album*, 114, in both the body text and the footnote on the same page.

96 Barthes, *Album*, 114.

97 Barthes, *Album*, xxiv.

98 Roland Barthes, *Mythologies*, trans. Richard Howard and Anette Lavers (New York: Hill and Wang, 2012 [1957]), 152.

99 Barthes, *Mythologies*, 160.

100 Barthes's essay fits within a growing disenchantment with information theory and its communication schema that unfolded in France starting in the late 1950s. On this history of importation and re-evaluation, see the essay originally published in 1966, Georges Mounin, "Communication, Linguistics and Information Theory" (1964), in *Semiotic Praxis: Studies in Pertinence and in the Means of Expression and Communication*, trans. Catherine Tihanyi, Maia Wise, and Bruce Wise (New York: Plenum, 1985), 33–49; and Algirdas Julien Greimas, "Semiotics and Social Communication" (1970 lecture), in *On Meaning: Selected Writings in Semiotic Theory*, trans. Paul J. Perron and Frank H. Collins (Minneapolis: University of Minnesota Press, 1987), 180–91. For remarks from a later participant in the Paris scene, see Francesco Casetti, *Semiotica: Saggio critico, testimonianze, documenti* (Milano: Accademia, 1977), 48–88.

101 Roland Barthes, "Le message photographique," *Communications*, no. 1 (1961): 127; my translation. See also Roland Barthes, "The Photographic Message" (1961), in *Image, Music, Text*, trans. Stephen Heath (New York: Hill and Wang, 1977), 15.

102 Barthes, "The Photographic Message," 15; translation modified. See the original in Roland Barthes, "Le message photographique," 127.

103 Barthes, "The Photographic Message," 31.

104 For more on Barthes's critique of informational economy, see Hayles, *Chaos Bound*, 187–96.

105 Barthes, *s/z*, 132.

106 Barthes, *s/z*, 131.

107 Barthes, *s/z*, 132.

108 Roudinesco, *Jacques Lacan*, 307.

109 Roland Barthes, "To the Seminar" (1974), in *The Rustle of Language*, trans. Richard Howard (Berkeley: University of California Press, 1989), 334, 341.

110 Barthes, "To the Seminar," 334, 337.

111 Barthes, *Album*, xxv.

112 Roland Barthes, "Writing the Event" (1968), in *The Rustle of Language*, trans. Richard Howard (Berkeley: University of California Press, 1989), 151.

113   Barthes, "Writing the Event," 151.

114   Barthes, "Writing the Event," 151.

115   Roland Barthes, *Sarrasine de Balzac: Séminaires à l'École pratique des hautes études, 1967–1968, 1968–1969* (Paris: Éditions du Seuil, 2011), 317; my translations (here and below).

116   Barthes, *Sarrasine de Balzac*, 317.

117   Barthes, *Sarrasine de Balzac*, 318.

118   Barthes, *Sarrasine de Balzac*, 319.

119   Barthes, *Sarrasine de Balzac*, 319.

120   Barthes, *Sarrasine de Balzac*, 319.

121   Barthes, *s/z*, 132.

122   Julia Kristeva, "Semiotics: A Critical Science and/or a Critique of Science" (1969), in *The Kristeva Reader*, ed. Toril Moi, trans. Seán Hand (New York: Columbia University Press, 1986), 74–89.

123   Luce Irigaray, "Linguistic and Specular Communication" (1966), in *To Speak Is Never Neutral*, trans. Gail Schwab (London: Continuum, 2002), 9–24.

124   Irigaray, "Linguistic and Specular Communication," 17,

125   Luce Irigaray, "Idiolect or Other Logic" (1985), in *To Speak Is Never Neutral*, 157.

126   For Irigaray's remarks on this nonrenewal, and its relation to the gendered norms governing the university, see Luce Irigaray, *Conversations* (London: Continuum, 2008), 74–75; on the University of Vincennes, see François Dosse, *History of Structuralism: The Sign Sets 1967–Present*, trans. Deborah Glassman (Minneapolis: University of Minnesota Press, 1997 [1991]), 141–53.

127   Luce Irigaray, "Introduction," in *To Speak Is Never Neutral*, 2.

128   Luce Irigaray, "Introduction," 4.

129   Irigaray, *To Speak Is Never Neutral*, 4.

130   *Communications*, "Activités du Centre d'études des communications de masse en 1968–1969," *Communications* 14, no. 1 (1969): 211; Jean Baudrillard, *The System of Objects*, trans. James Benedict (London: Verso, 1996 [1968]), 127.

131   Jean Baudrillard, "Requiem for the Media" (1972), in *For a Critique of the Political Economy of the Sign*, trans. Charles Levin (St. Louis, MO: Telos Press, 1981), 178.

132   Baudrillard, "Requiem for the Media," 178. Italics in the original.

133   For more on these dynamics, see Pierre Bourdieu and Loïc Wacquant, "On the Cunning of Imperialist Reason," *Theory, Culture and Society* 16, no. 1 (1999): 41–58, https://doi.org/10.1177/026327699016001003.

134   On Derrida, cybernetics, and transparency, see Stefanos Geroulanos, *Transparency in Postwar France: A Critical History of the Present* (Stanford, CA: Stanford University Press, 2017), 267–331.

Conclusion

1   Anna Wiener, *Uncanny Valley: A Memoir* (London: 4th Estate, 2020), 71.

2   Wiener, *Uncanny Valley*, 41.

3   Wiener, *Uncanny Valley*, 43.

4   On the historical links of Bay Area counterculture to informatics, see Richard Barbrook and Andy Cameron, "The Californian Ideology," *Science as Culture* 6, no. 1 (1996): 44–72, https://doi.org/10.1080/09505439609526455; and Fred Turner, *From Counterculture to Cyberculture: Stewart Brand, the Whole Earth Network, and the Rise of Digital Utopianism* (Chicago: University of Chicago Press, 2006). See also Lee, *Think Tank Aesthetics*.

5   See, for example, Lilly Irani, "Difference and Dependence among Digital Workers: The Case of Amazon Mechanical Turk," *South Atlantic Quarterly* 114, no. 1 (Winter 2015): 225–34; Virginia Eubanks, *Automating Inequality: How High-Tech Tools Profile, Police, and Punish the Poor* (New York: St. Martin's, 2018); Jackie Wang, *Carceral Capitalism* (South Pasadena, CA: Semiotext(e), 2018); Ames, *The Charisma Machine*; Darren Byler, "The Digital Enclosure of Turkic Muslims," *Society + Space*, December 7, 2020, https://www.societyandspace.org/articles/the-digital-enclosure-of-turkic-muslims.

6   There are notable exceptions to this trend, such as Andrew Piper's often sweeping overviews of sources for statistical and other styles of computational reason in Andrew Piper, *Can We Be Wrong? The Problem of Textual Evidence in a Time of Data Elements in Digital Literary Studies* (Cambridge: Cambridge University Press, 2020). Nonetheless, it seems that the call for "future large-scale studies of disciplinary behavior" (7) would benefit from consideration of the human sciences properly, with attention to work in fields such as colonial ethnography, psychotherapy, national infrastructure policy, and penal education, which the foregoing chapters suggest may prove worthy of consideration as thoughtful and nuanced as that given by Piper to Laplace and Kant. Fortunately, this is not an either-or situation, but rather an occasion for composition.

7   Chris Anderson, "The End of Theory: The Data Deluge Makes the Scientific Method Obsolete," *Wired*, June 23, 2008, https://www.wired.com/2008/06/pb-theory/.

8   Franco Moretti, *Graphs, Maps, Trees: Abstract Models for a Literary History* (London: Verso, 2005), 2.

9   Franco Moretti, "'Operationalizing': Or, the Function of Measurement in Literary Theory," *New Left Review* 84 (December 2013): 103–19.

10   Moretti, "Operationalizing," 103–4.

11   Lacan, "The Freudian Thing," 350.

12   Because such extensive discussion has been given to these other scholars earlier in the book, here I will only briefly recall Warren Weaver's fascinating "Statistical Freedom of the Will."

13   To briefly recall notes from earlier chapters, see Lacan, *The Ego in Freud's Theory*, 175–273.

14   Judith Butler, *Gender Trouble: Feminism and the Subversion of Identity* (New York: Routledge, 1990), 35.

15   Paul Michael Kurtz, "The Philological Apparatus: Science, Text, and Nation in the Nineteenth Century," *Critical Inquiry* 47, no. 4 (Summer 2021): 747–76, https://doi.org/10.1086/714541.

16   Mead, "Cybernetics of Cybernetics," 2.

17   Piper, *Can We Be Wrong*, 19.

18   Katherine Bode, *Reading by Numbers: Recalibrating the Literary Field* (London: Anthem, 2012), 10.

19   Katie Trumpener, "Critical Response I. Paratext and Genre System: A Response to Franco Moretti," *Critical Inquiry* 36, no. 1 (Autumn 2009): 159–71, https://doi.org/10.1086/606126.

20   Ted Underwood, *Distant Horizons: Digital Evidence and Literary Change* (Chicago: University of Chicago Press, 2019), 167.

21   Bode, *Reading by Numbers*, 11–12.

22   With this phrasing I have in mind historian Rebecca Lemov's characterization of contemporaneous initiatives at Harvard's Institute of Social Relations, which tried, without success, to recruit Lévi-Strauss for a full professorship and occasionally consulted with Jakobson.

23   Lacan, *The Ego*, 47. Translation modified.

24   Martin Heidegger, "The Question Concerning Technology" (1955 lecture), in *The Question Concerning Technology and Other Essays* (New York: Harper and Row, 1977), 28.

25   Michael Dudley Johnson and Ryan Patterson, Suggesting search results to users before receiving any search query from the users, United States US10467239B2, filed June 8, 2018, and issued November 5, 2019, https://patents.google.com/patent/US10467239B2/en.

Bibliography

A note on dates: Where possible, I have indicated the date of a work's initial presentation (usually the first print publication date but occasionally the first lecture date). Where the work first appeared in one form and then subsequently appeared in a compilation, I include the original date after the title of text rather than after the title of the book in which it appeared. This approach seeks to balance the practical challenge posed by the fact that key compilations and their translation, such as Lacan's *Écrits* or Bateson's *Ecology of Mind*, have their own historical significance. The present approach aims at highlighting when an initial text or statement entered into public discourse, while also acknowledging the volumes and translations that also entered into these works' developing historical significance.

Archival Sources and Abbreviations

ELP   École Libre Papers, New School for Social Research Archives, New York

FBIA   Federal Bureau of Investigation Archives, Washington, DC

JA   Don D. Jackson Archive, University of Louisiana, Monroe, LA

JBW   J. B. Wiesner Papers, MIT Distinctive Collections, Cambridge, MA

MM   Margaret Mead Papers and South Pacific Ethnographic Archives, 1838–1996, Manuscript Division, Library of Congress, Washington, DC

NWP   Norbert Wiener Papers, MIT Distinctive Collections, Cambridge, MA

NYPL   New York Public Library, New York

RAC   Rockefeller Archive Center, Sleepy Hollow, NY

RBP   Ray L. Birdwhistell Papers, Folklore and Ethnography Archives, University of Pennsylvania Philadelphia, PA

RF   Rockefeller Foundation Archives, Sleepy Hollow, NY (subset of RAC)

RJP   Roman Jakobson Papers, MIT Distinctive Collections, Cambridge, MA

All Other Sources

Alder, Ken. *Engineering the Revolution: Arms and Enlightenment in France, 1763–1815.* Princeton, NJ: Princeton University Press, 1997.

Alliez, Eric, and Maurizio Lazzarato. *Wars and Capital*. Translated by Ames Hodges. South Pasadena, CA: Semiotext(e), 2016.

Althusser, Louis. "Philosophy and Social Science: Introducing Bourdieu and Passeron" (1963 lecture). Translated by Rachel Gomme. *Theory, Culture & Society* 36, no. 7–8 (2019): 5–21.

Althusser, Louis. *Psychoanalysis and the Human Sciences* (lectures in 1963–64). Translated by Steven Rendall. New York: Columbia University Press, 2016.

American Telephone and Telegraph Company. "Our Many-Tongued Ancestors." *Survey* 43, no. 19 (March 6, 1920): 715.

Ames, Morgan. *The Charisma Machine: The Life, Death, and Legacy of One Laptop per Child*. Cambridge, MA: MIT Press, 2019.

Anderson, Chris. "The End of Theory: The Data Deluge Makes the Scientific Method Obsolete." *Wired*, June 23, 2008. https://www.wired.com/2008/06/pb-theory/.

Angermuller, Johannes. *Why There Is No Poststructuralism in France: The Making of an Intellectual Generation*. London: Bloomsbury Academic, 2015.

Asaro, Peter. "McCulloch, Warren S. (Artificial Neural Networks)." Peterasaro.org. Accessed December 3, 2020. https://peterasaro.org/writing/McCulloch.html.

Barbrook, Richard, and Andy Cameron. "The Californian Ideology." *Science as Culture* 6, no. 1 (1996): 44–72. https://doi.org/10.1080/09505439609526455.

Baring, Edward. "Liberalism and the Algerian War: The Case of Jacques Derrida." *Critical Inquiry* 36, no. 2 (Winter 2010): 239–61. https://doi.org/10.1086/648525.

Barthes, Roland. *Album: Unpublished Correspondence and Texts*. Edited by Éric Marty. Translated by Jody Gladding. New York: Columbia University Press, 2018.

Barthes, Roland. "Le message photographique." *Communications*, no. 1 (1961): 127–38.

Barthes, Roland. *Mythologies*. Translated by Richard Howard and Anette Lavers. 1957. New York: Hill and Wang, 2012.

Barthes, Roland. "The Photographic Message" (1961). In *Image, Music, Text*, translated by Stephen Heath, 15–31. New York: Hill and Wang, 1977.

Barthes, Roland. *Sarrasine de Balzac: Séminaires à l'École pratique des hautes études, 1967–1968, 1968–1969*. 1967–69. Paris: Éditions du Seuil, 2011.

Barthes, Roland. *s/z: An Essay*. Translated by Richard Miller. 1973. Reprint, New York: Farrar, Straus and Giroux, 1974.

Barthes, Roland. "To the Seminar" (1974). In *The Rustle of Language*, translated by Richard Howard, 332–42. Berkeley: University of California Press, 1989.

Barthes, Roland. "Writing the Event" (1968). In *The Rustle of Language*, translated by Richard Howard, 149–54. Berkeley: University of California Press, 1989.

Bateson, Gregory. "Atoms, Nations, and Cultures." *International House Quarterly* 11, no. 2 (1947): 47–50.

Bateson, Gregory. "Epidemiology of a Schizophrenia" (1955). In Bateson, *Steps to an Ecology of Mind*, 194–200.

Bateson, Gregory. "Minimal Requirements for a Theory of Schizophrenia" (1960). In Bateson, *Steps to an Ecology of Mind*, 244–70.

Bateson, Gregory. *Naven: A Survey of the Problems Suggested by a Composite Picture of the Culture of a New Guinea Tribe Drawn from Three Points of View*. 1936. Stanford, CA: Stanford University Press, 1958.

Bateson, Gregory. "Notes on the Photographs and Captions." In Bateson and Mead, *Balinese Character*, 49–54.

Bateson, Gregory. "Social Planning and the Concept of Deutero-Learning" (1942). In Bateson, *Steps to an Ecology of Mind*, 159–76.

Bateson, Gregory. *Steps to an Ecology of Mind*. New York: Ballantine, 1972.

Bateson, Gregory, Don D. Jackson, Jay Haley, and John Weakland. "Toward a Theory of Schizophrenia." *Behavioral Science* 1, no. 4 (1956): 251–64.

Bateson, Gregory, and Weldon Kees. "The Nature of Play: Part 1, River Otters" (ca. 1954). *Grey Room*, no. 66 (Winter 2017): 110–14. https://doi.org/10.1162/GREY_a_00215.

Bateson, Gregory, and Margaret Mead. *Balinese Character: A Photographic Analysis*. New York: New York Academy of Sciences, 1942.

Bateson, Mary Catherine. "Continuities in Insight and Innovation: Toward a Biography of Margaret Mead." *American Anthropologist* 82, no. 2 (1980): 270–77.

Bateson, Mary Catherine. *With a Daughter's Eye: A Memoir of Margaret Mead and Gregory Bateson*. New York: William Morrow, 1984.

Baudrillard, Jean. "Requiem for the Media" (1972). In *For a Critique of the Political Economy of the Sign*, translated by Charles Levin, 164–84. St. Louis, MO: Telos, 1981.

Baudrillard, Jean. *The System of Objects*. Translated by James Benedict. 1968. London: Verso, 1996.

Bauer, Raymond A., Alex Inkeles, and Clyde Kluckhohn. *How the Soviet System Works: Cultural, Psychological, and Social Themes*. Cambridge, MA: Harvard University Press, 1956.

Beauvoir, Simone de. *Les belles images*. Translated by Patrick O'Brian. 1966. Reprint, London: Fontana/Collins, 1975.

Belcher, Oliver. "Sensing, Territory, Population: Computation, Embodied Sensors, and Hamlet Control in the Vietnam War." *Security Dialogue* 50, no. 5 (2019): 416–36. https://doi.org/10.1177/0967010619862447.

Belhoste, Bruno. *La formation d'une technocratie: L'école polytechnique et ses élèves de la révolution au second empire*. Paris: Belin, 2003.

Bell, Daniel. "Notes on the Post-Industrial Society: Part I." *Public Interest*, no. 6 (1967): 24–35.

Benedict, Ruth. *Patterns of Culture*. 1934. Boston: Houghton Mifflin, 1959.

Benjamin, Walter. "The Work of Art in the Age of Its Technological Reproducibility" (first version; 1935). Translated by Michael W. Jennings. *Grey Room*, no. 39 (Spring 2010): 11–37.

Berman, Edward H. *The Ideology of Philanthropy: The Influence of the Carnegie, Ford, and Rockefeller Foundations on American Foreign Policy*. Albany, NY: SUNY Press, 1983.

Bode, Katherine. *Reading by Numbers: Recalibrating the Literary Field*. London: Anthem, 2012.

Boltanski, Luc. "Visions of American Management in Post-War France." Translated by Alexandra Russell. *Theory and Society* 12, no. 3 (1983): 375–403.

Boole, George. *An Investigation of the Laws of Thought, on Which Are Founded the Mathematical Theories of Logic and Probabilities*. London: Macmillan, 1854.

Bourdieu, Pierre. "Préface." In *Aux origines de l'école des hautes études en sciences sociales: Le rôle du mécénat Américain (1920–1960)*, i–iv. Paris: Les éditions du Cerf, 1988.

Bourdieu, Pierre. *The State Nobility: Elite Schools in the Field of Power*. Translated by Lauretta C. Clough. 1989. Cambridge: Polity, 1996.

Bourdieu, Pierre, and Loïc Wacquant. "On the Cunning of Imperialist Reason." *Theory, Culture and Society* 16, no. 1 (1999): 41–58. https://doi.org/10.1177/026327699016001003.

Bowker, Geoffrey. "How to Be Universal: Some Cybernetic Strategies, 1943–70." *Social Studies of Science* 23, no. 1 (1993): 107–27.

Brahnam, S. Berlin. "On Lacan's Neglected Computational Model and the Oedipal Structure." *S: Journal of the Circle for Lacanian Ideology Critique*, nos. 10–11 (2017–18): 261–74.

Brahnam, S. Berlin. "Primordia of Après-Coup, Fractal Memory, and Hidden Letters." *S: Journal of the Circle for Lacanian Ideology Critique*, nos. 10–11 (2017–18): 202–44.

Braidotti, Rosi. *The Posthuman*. Cambridge: Polity, 2013.

Brain, Robert. "Standards and Semiotics." In *Inscribing Science: Scientific Texts and the Materiality of Communication*, edited by Timothy Lenoir, 249–84. Stanford, CA: Stanford University Press, 1998.

Brand, Stewart. *II Cybernetic Frontiers*. New York: Random House, 1974.

Bryson, Dennis. "Lawrence K. Frank, Knowledge, and the Production of the 'Social.'" *Poetics Today* 19, No. 3 (Autumn, 1998): 401–21.

Buck-Morss, Susan. "Aesthetics and Anaesthetics: Walter Benjamin's Artwork Essay Reconsidered." *October*, no. 62 (1992): 3–41. https://doi.org/10.2307/778700.

Burke, Colin. *Information and Secrecy: Vannevar Bush, Ultra, and the Other Memex*. Metuchen, NJ: Scarecrow, 1994.

Bush, Vannevar. "As We May Think." *Atlantic Monthly*, July 1945.

Bush, Vannevar. "As We May Think." *Life*, September 1945.

Butler, Judith. *Gender Trouble: Feminism and the Subversion of Identity*. New York: Routledge, 1990.

Buxton, William. "From Radio Research to Communication Intelligence: Rockefeller Philanthropy, Communications Specialists, and the American Policy Community." In *Communication Researchers and Policy-Making*, edited by Sandra Braman, 295–346. Cambridge, MA: MIT Press, 2003.

Buxton, William. "Rockefeller Philanthropy and Communications, 1935–1939." In *The Development of the Social Sciences in the United States and Canada: The Role of Philanthropy*, edited by Theresa Richardson and Donald Fisher, 177–92. Stamford, CT: Ablex, 1999.

Byler, Darren. "The Digital Enclosure of Turkic Muslims." *Society + Space*, December 7, 2020. https://www.societyandspace.org/articles/the-digital-enclosure-of -turkic-muslims.

Campolo, Alexander. "Signs and Sight: Jacques Bertin and the Visual Language of Structuralism." *Grey Room*, no. 78 (Winter 2020): 34–65. https://doi.org/10 .1162/grey_a_00286.

Canguilhem, Georges. "Machine and Organism" (1947 lecture). Translated by Mark Cohen and Randall Cherry. In *Incorporations*, edited by Jonathan Crary and Sanford Kwinter, 45–69. New York: Zone, 1992.

Canguilhem, Georges. "What Is Psychology?" (1958). Translated by David M. Peña-Guzmán. *Foucault Studies*, no. 21 (June 2016): 200–213.

Carey, James W. "A Cultural Approach to Communications" (1989). In *Communication as Culture: Essays on Media and Society*, 11–28. New York: Routledge, 2009.

Carey, James W. "Technology and Ideology: The Case of the Telegraph" (1983). In *Communication as Culture: Essays on Media and Society*, 155–77. New York: Routledge, 2009.

Casetti, Francesco. *Semiotica: Saggio critico, testimonianze, documenti*. Milano: Accademia, 1977.

Chambarlhac, Vincent. "Lévi-Strauss en socialisme." *Cahiers d'histoire: Revue d'histoire critique*, no. 101 (2007): 83–99.

Chaubet, François and Emmanuelle Loyer. "L'École libre des hautes études de New York: Exil et résistance intellectuelle (1942–1946)." *Revue Historique* 302, no. 4/616 (October–December 2000): 939–72.

Cherry, E. Colin. "A History of the Theory of Information." *Proceedings of the IEE Part III: Radio and Communication Engineering* 98, no. 55 (September 1951): 383–93. https://doi.org/10.1049/pi-3.1951.0082.

Cherry, E. Colin, Morris Halle, and Roman Jakobson. "Toward the Logical Description of Languages in Their Phonemic Aspect." *Language* 29, no. 1 (March 1953): 34–46.

Chiu, Eugen, Jocelyn Lin, Brok Mcferron, Noshirwan Petigara, and Satwiksai Seshasai. "Mathematical Theory of Claude Shannon: A Study of the Style and Context of His Work up to the Genesis of Information Theory." MIT, Cambridge, MA, 2001. https://web.mit.edu/6.933/www/Fall2001/Shannon1.pdf.

Chomsky, Noam. *Language and Mind*. 1968. Cambridge: Cambridge University Press, 2006.

Collopy, Peter Sachs. "The Revolution Will Be Videotaped: Marking a Technology of Consciousness in the Long 1960s." PhD diss., University of Pennsylvania, 2015.

Colomina, Beatriz. "Enclosed by Images: The Eameses' Multimedia Architecture." *Grey Room*, no. 2 (Winter 2001): 7–29. https://www.jstor.org/stable/1262540.

Communications. "Activités du Centre d'études des communications de masse en 1968–1969." *Communications* 14, no. 1 (1969): 211–15 https://www.persee.fr /docAsPDF/comm_0588-8018_1969_num_14_1_1210.

Communications. "Vie des centres 1960–1961." *Communications* 1, no. 1 (1961): 226–30. http://www.persee.fr/web/revues/home/prescript/article/comm_0588-8018 _1961_num_1_1_939.

Copjec, Joan. *Read My Desire: Lacan against the Historicists*. Cambridge, MA: MIT Press, 1994.

Culbert, David. "The Rockefeller Foundation, the Museum of Modern Art Film Library, and Siegfried Kracauer, 1941." *Historical Journal of Film, Radio and Television* 13, no. 4 (1993): 495–511.

Cusset, François. "Cybernétique et « théorie française »: Faux alliés, vraies ennemis." *Multitudes Web* 22 (Autumn 2005): https://www.cairn.info/revue-multitudes-2005-3-page-223.htm.

Cusset, François. *French Theory: How Foucault, Derrida, Deleuze, and Co. Transformed the Intellectual Life of the United States*. Translated by Jeff Fort, Josephine Berganza, and Marlon Jones. 2003. Minneapolis: University of Minnesota Press, 2008.

Daston, Lorraine, and Peter Galison. "The Image of Objectivity." *Representations*, no. 40 (Autumn 1992): 81–128.

de Man, Paul. "Roland Barthes and the Limits of Structuralism" (ca. 1972 composition). *Yale French Studies*, no. 77 (1990): 177–90. https://doi.org/10.2307/2930153.

Derrida, Jacques. "Force and Signification" (1963 lecture, 1967 print). In *Writing and Difference*, translated by Alan Bass, 3–30. Chicago: University of Chicago Press, 1978.

Derrida, Jacques. "Implications: Interview with Henry Ronse" (1967). In *Positions*, translated by Alan Bass, 1–14. Chicago: University of Chicago Press, 1981.

Derrida, Jacques. *La vie la mort: Séminaire (1975–1976)*. Edited by Pascale-Anne Brault and Peggy Kamuf. Paris: Éditions du Seuil, 2019.

Derrida, Jacques. *Of Grammatology*. Translated by Gayatri Spivak. 1967. Baltimore, MD: Johns Hopkins University Press, 1976.

Descombes, Vincent. *Modern French Philosophy*. 1979. Cambridge: Cambridge University Press, 1980.

Deutsch, Albert. *The Shame of the States*. New York: Harcourt, Brace, 1948.

DiLeonardi, Sean Michael. "Cryptographic Reading: Machine Translation, the New Criticism, and Nabokov's *Pnin*." *Post45*, January 2019. https://post45.org /2019/01/cryptographic-reading-machine-translation-the-new-criticism-and -nabokovs-pnin/.

Dosse, François. *History of Structuralism: The Rising Sign, 1945–1966*. Translated by Deborah Glassman. 1991. Minneapolis: University of Minnesota Press, 1998.

Dosse, François. *History of Structuralism: The Sign Sets, 1967–Present*. Translated by Deborah Glassman. 1992. Minneapolis: University of Minnesota Press, 1997.

Douglas, Mary. "Preface." In Mauss, *The Gift: The Form and Reason for Exchange in Archaic Societies*, ix–xxiii. New York: Routledge, 1990.

Dowie, Mark. *American Foundations: An Investigative History*. Cambridge, MA: MIT Press, 2001.

Dreyfus, Dina. "De la pensée prelogique à la pensée hyperlogique (à propos de la pensée sauvage)." *Mercure de France*, February 1963: 309–22.

Dreyfus, Dina, Alain Badiou, Jean Hyppolite, Georges Mounin, Pierre Bourdieu, and Jean Laplanche. *Emissions de philosophie 1966–1967 consacrées au langage*. Paris: Institut pédagogique, Dossiers pédagogiques de la radiotélevision scolaire, 1967.

Dubarle, Père Dominique. "Vers la machine à gouverner." *Le Monde*, December 28, 1948: 47–49.

Dumont, Lucile. "From Sociology to Literary Theory: The Disciplinary Affiliations of Literature in Section VI of the École Pratique des Hautes Études (EPHE), 1956–1975." Translated by Jean-Yves Bart. *Symbolic Goods* 3 (2018). https://www.biens-symboliques.net/289?lang=en.

Dumont, Lucile. "Literary Theorists in and beyond French Academic Space (1960–1970s)." *Sociological Review* 68, no. 5 (2020): 1108–23. https://doi.org/10.1177/0038026120916119.

Duncan, Dennis. *The Oulipo and Modern Thought*. Oxford: Oxford University Press, 2019.

Edwards, Paul N. *The Closed World: Computers and the Politics of Discourse in Cold War America*. Cambridge, MA: MIT Press, 1996.

Eggan, Fred, William C. Boyd, S. L. Washburn, Clyde Kluckhohn, George P. Murdock, Jules Henry, Harry Hoijer, Edward A. Kennard, André Martinet, Claude Lévi-Strauss, Margaret Mead, Gordon Willey, Wendell C. Bennett, A. L. Kroeber, S. F. Nadel, Meyer Schapiro, J. O. Brew, Joseph H. Greenberg, Alfonso Caso, Ralph Linton, Leslie A. White, Robert Redfield, Daryll Forde, Julian H. Steward, Oscar Lewis, Ralph Beals, and John Howland Rowe. "Problems of Process: Results" (roundtable discussion). In *An Appraisal of Anthropology Today*, edited by Sol Tax, Loren C. Eiseley, Irving Rouse, and Carl F. Voegelin, 104–24. Chicago: University of Chicago Press, 1953.

Engelke, Henning. "Filmisches Wissen und der Geist des Kalten Krieges: Kybernetische Modelle bei Gregory Bateson und Weldon Kees." In *Wissensraum Film*, edited by Irina Gradinari, Dorit Müller, and Johannes Pause, 225–41. Wiesbaden: Reichert Verlag, 2014.

Erickson, Paul, Judy L. Klein, Lorraine Daston, Rebecca Lemov, Thomas Sturm, and Michael D. Gordin. *How Reason Almost Lost Its Mind: The Strange Career of Cold War Rationality*. Chicago: University of Chicago Press, 2013.

Eubanks, Virginia. *Automating Inequality: How High-Tech Tools Profile, Police, and Punish the Poor*. New York: St. Martin's, 2018.

Fant, Gunnar. "Speech Research in a Historical Perspective" (alternate title: "Phonetics and Phonology in the Last 50 Years"). Paper presented at From Sound to Sense, MIT, Research Laboratory of Electronics, June 11–13, 2004. https://www.rle.mit.edu/soundtosense/conference/pdfs/invitedspeakers/Fant%20PAPER.pdf.

Ford, Clellan S. *Human Relations Area Files: 1959–1969. A Twenty-Year Report*. New Haven, CT: Human Relations Area Files, 1970.

Fosdick, Raymond B. *The Old Savage in the New Civilization*. New York: Doubleday, Doran, 1929.

Fosdick, Raymond B. *The Story of the Rockefeller Foundation*. New York: Harper, 1952.

Foucault, Michel. "Linguistique et sciences sociales" (1969 lecture). In *Dits et écrits, tome 1*, 821–42. Paris: Editions Gallimard, 1994.

Foucault, Michel. "Message ou bruit" (1966). In *Dits et écrits, tome 1*, 557–60. Paris: Editions Gallimard, 1994.

Foucault, Michel. "Truth and Power" (1977). In *Power/Knowledge: Selected Interviews and Other Writings*, edited by Colin Gordon, 109–33. New York: Pantheon Books, 1980.

Foucault, Michel, and Alain Badiou. "Philosophie et psychologie" (1965 broadcast). In *Dits et écrits, tome 1*, 438–48. Paris: Editions Gallimard, 1994.

Fougeyrollas, Pierre. "Bureaucratie et technocratie." *Arguments* 4, no. 17 (1960): 31–32.

Fournier, Marcel. *Marcel Mauss: A Biography*. Translated by Jane Marie Todd. Princeton, NJ: Princeton University Press, 2006 (1994).

Frank, Lawrence K. "The Family as Cultural Agent." *Living* 2, no. 1 (1940): 16–19. https://doi.org/10.2307/346505.

Franklin, Seb. "Commodities, Love, and Attenuated Reproduction in the Archives of Cybernetics." *Diacritics* (blog), June 3, 2021. https://www.diacriticsjournal .com/unarchive-6/.

Fremont-Smith, Frank. "Introductory Discussion," (1949 address). In Pias, *Cybernetics*, 29–40.

Funck-Brentano, Theophile. *Les sciences humaines: Philosophie, médicine, morale, politique*. Paris: Librairie internationale, 1868.

Galison, Peter. "The Americanization of Unity." *Daedalus* 127, no. 1 (1998): 45–71.

Galison, Peter. "The Ontology of the Enemy: Norbert Wiener and the Cybernetic Vision." *Critical Inquiry* 21, no. 1 (Autumn 1994): 228–68.

Galton, Francis. "Arithmetic Notation of Kinship." *Nature* 28, no. 723 (September 6, 1883): 435. https://doi.org/10.1038/028435b0.

Gary, Brett. "Communication Research, the Rockefeller Foundation, and Mobilization for the War on Words, 1938–1944." *Journal of Communication* 46, no. 3 (1996): 124–48.

Geertz, Clifford. "The Cerebral Savage: On the Work of Claude Lévi-Strauss" (1967). In *The Interpretation of Cultures: Selected Essays*, 345–59. New York: Basic Books, 1973.

Gemelli, Giuliana. *The Ford Foundation and Europe*. Brussels: European Interuniversity Press, 1998.

Gemelli, Giuliana, ed. *The "Unacceptables": American Foundations and Refugee Scholars between the Two Wars and After*. Brussels: Peter Lang P.I.E., 2000.

Geoghegan, Bernard Dionysius. "Editor's Introduction: What Bound the Double Bind?" *Grey Room*, no. 66 (Winter 2017): 102–9. https://doi.org/10.1162/GREY_a _00213.

Geroulanos, Stefanos. *Transparency in Postwar France: A Critical History of the Present.* Stanford, CA: Stanford University Press, 2017.

Geroulanos, Stefanos. "Violence and Mechanism: Georges Canguilhem's Overturning of the Cartesian Legacy." *Qui Parle* 24, no. 1 (Fall/Winter 2015): 125–46. https://doi.org/10.5250/quiparle.24.1.0125.

Gerovitch, Slava. "Roman Jakobson und die Kybernetisierung der Linguistik in der Sowjetunion." In *Die Transformation des Humanen: Beiträge zur Kulturgeschichte der Kybernetik*, edited by Erich Hörl and Michael Hagner, 229–74. Frankfurt: Suhrkamp, 2008.

Goffman, Erving. *Asylums: Essays on the Social Situation of Mental Patients and Other Inmates.* Garden City, NJ: Anchor Books, 1961.

Goleman, Daniel. "Deadlines for Change: Therapy in the Age of Reaganomics." *Psychology Today*, August 1981.

Good, Paul R. "Brief Therapy in the Age of Reagapeutics." *American Journal of Orthopsychiatry* 57, no. 1 (January 1987): 6–11.

Gray, Chris Hables, ed. *The Cyborg Handbook.* London: Routledge, 1995.

Greimas, Algirdas Julien. "Semiotics and Social Communication" (1970). In *On Meaning: Selected Writings in Semiotic Theory*, translated by Paul J. Perron and Frank H. Collins, 180–91. Minneapolis: University of Minnesota Press, 1987.

Gurvitch, Georges. "Le concept de structure sociale." *Cahiers internationaux de sociologie* 19 (July–December 1955): 3–44.

Habermas, Jürgen. "Technology and Science as 'Ideology'" (1968). Translated by Jeremy J. Shapiro. In *Toward a Rational Society*, 81–127. Boston: Beacon, 1970.

Habermas, Jürgen. *The Theory of Communicative Action, v. I: Reason and the Rationalization of Society.* Translated by Thomas McCarthy. Boston: Beacon Press, 1984 [1981].

Hagelbarger, D. W. "SEER, A SEquence Extrapolating Robot." *IRE Transactions on Electronic Computers* EC-5, no. 1 (March 1956): 1–4.

Haley, Jay. "Development of a Theory: A History of a Research Project." In *Double Bind: The Foundation of the Communicational Approach to the Family*, edited by Carlos E. Sluski and Donald C. Ransom, 59–104. New York: Grune and Stratton, 1976.

Haley, Jay. "Family Experiments: A New Type of Experimentation." *Family Process* 1, no. 2 (September 1962): 265–93.

Hall, Stuart. "Cultural Studies and the Centre: Some Problematics and Problems." In *Culture, Media, Language: Working Papers in Cultural Studies, 1972–79*, edited by Stuart Hall, 15–47. London: Hutchinson and Centre for Contemporary Cultural Studies, University of Birmingham, 1980.

Halpern, Joseph. *The Myths of Deinstitutionalization: Policies for the Mentally Disabled.* Boulder, CO: Westview, 1980.

Hansen, Mark B. N. *New Philosophy for New Media.* Cambridge, MA: MIT Press, 2004.

Haraway, Donna J. "The Biological Enterprise: Sex, Mind, and Profit from Human Engineering to Sociobiology." *Radical History Review* 20 (Spring/Summer 1979): 206–37.

Haraway, Donna J. "The High Cost of Information in Post–World War II Evolutionary Biology: Ergonomics, Semiotics, and the Sociobiology of Communications Systems." *Philosophical Forum* 13, nos. 2–3 (1981–82): 244–78.

Haraway, Donna J. "Manifesto for Cyborgs: Science, Technology, and Socialist Feminism in the 1980s." *Socialist Review*, no. 80 (1985): 65–108.

Haraway, Donna J. "Signs of Dominance: From a Physiology to a Cybernetics of Primate Society, C. R. Carpenter, 1930–1970." In *Studies in History of Biology*, edited by William R. Coleman and Camille Limoges, 129–219. Baltimore, MD: Johns Hopkins University Press, 1983.

Hayles, N. Katherine. *Chaos Bound: Orderly Disorder in Contemporary Literature and Science*. Ithaca, NY: Cornell University Press, 1990.

Hayles, N. Katherine. *How We Became Posthuman: Virtual Bodies in Cybernetics, Literature, and Informatics*. Chicago: University of Chicago Press, 1999.

Hecht, Gabrielle. *The Radiance of France: Nuclear Power and National Identity after World War II*. Cambridge, MA: MIT Press, 2009.

Heidegger, Martin. "The Question concerning Technology" (1955 lecture). In *The Question concerning Technology and Other Essays*, 3–35. New York: Harper and Row, 1977.

Heims, Steve J. *Constructing a Social Science for Postwar America: The Cybernetics Group (1946–1953)*. Cambridge, MA: MIT Press, 1991.

Herman, Ellen. *The Romance of American Psychology: Political Culture in the Age of Experts*. Berkeley: University of California Press, 1995.

Heyck, Hunter. *Age of System: Understanding the Development of Modern Social Science*. Baltimore, MD: Johns Hopkins University Press, 2015.

Hillyer, Minette. "Camera Documents Made at Home: Visual Culture and the Question of America." *Film History* 27, no. 4 (2015): 46–75.

Horkheimer, Max. "Traditional and Critical Theory" (1937). In *Critical Theory: Selected Essays*, translated by Matthew J. O'Connell, 188–243. New York: Continuum, 1972.

Howard, Jane. *Margaret Mead: A Life*. New York: Ballantine Books, 1980.

Hutchins, W. John. "Warren Weaver and the Launching of MT." In *Early Years in Machine Translation: Memoirs and Biographies of Pioneers*, edited by W. John Hutchins, 17–20. Philadelphia: John Benjamins, 2000.

Inventory of the Gregory Bateson Papers (MS 98), Special Collections and Archives, University of California, Santa Cruz, 2008. http://oac.cdlib.org/findaid/ark:/13030/kt029029gz.

Irani, Lilly. "Difference and Dependence among Digital Workers: The Case of Amazon Mechanical Turk." *South Atlantic Quarterly* 114, no. 1 (Winter 2015): 225–34.

Irigaray, Luce. *Conversations*. London: Continuum, 2008.

Irigaray, Luce. "Idiolect or Other Logic" (1985). In Irigaray, *To Speak Is Never Neutral*, 153–72.

Irigaray, Luce. "Introduction." In Irigaray, *To Speak Is Never Neutral*, 1–7.

Irigaray, Luce. "Linguistic and Specular Communication" (1966). In Irigaray, *To Speak Is Never Neutral*, 9–24.

Irigaray, Luce. *To Speak Is Never Neutral*. Translated by Gail Schwab. London: Continuum, 2002 (1985 essay collection).

Jacknis, Ira. "Margaret Mead and Gregory Bateson in Bali: Their Use of Photography and Film." *Cultural Anthropology* 3, no. 2 (May 1988): 160–77.

Jackson, Don D. "The Eternal Triangle: An Interview with Don D. Jackson, M.D." In *Techniques of Family Therapy*, edited by Jay Haley and L. Hoffman. New York: Basic Books, 1967.

Jakobson, Roman. "Closing Statements: Linguistics and Poetics" (1958 lecture). In *Style in Language*, ed. Thomas A. Sebeok, 350–77. Cambridge, MA: MIT Press, 1960.

Jakobson, Roman. "Memoir" (recorded in 1977). In Jakobson, *My Futurist Years*, 3–100.

Jakobson, Roman. *My Futurist Years*. Edited by Bengt Jangfeldt, translated by Stephen Rudy. New York: Marsilio, 1997.

Jakobson, Roman. "The Generation That Squandered Its Poets (Excerpts)" (1930). Translated by Dale E. Peterson. *Yale French Studies*, no. 39 (1967): 119–25.

Jakobson, Roman. "Principles of Historical Phonology" (1949 revision). In *On Language*, edited by Linda R. Waugh and Monique Monville-Burston, 184–201. Cambridge, MA: Harvard University Press, 1990.

Jakobson, Roman. "Results of a Joint Conference of Anthropologists and Linguists" (1952 lecture). In *Selected Writings II: Word and Language*, 554–67. The Hague: Mouton, 1971.

Jakobson, Roman. *Six Lectures on Sound and Meaning* (lectures ca. 1942, revised publication 1976). Cambridge, MA: MIT Press, 1978.

Jakobson, Roman. "On a Generation That Squandered Its Poets" (1930). In Jakobson, *My Futurist Years*, 209–45.

Jakobson, Roman, and Morris Halle. *Fundamentals of Language*. The Hague: Mouton, 1956.

Jakobson, Roman, and Claude Lévi-Strauss. *Correspondance 1942–1982*. Edited by Emmanuelle Loyer and Patrice Maniglier. Paris: Éditions du Seuil, 2018.

Johnson, Christopher. *Claude Lévi-Strauss: The Formative Years*. New York: Cambridge University Press, 2003.

Johnson, Michael Dudley, and Ryan Patterson. "Suggesting Search Results to Users before Receiving Any Search Query from the Users." United States US10467239B2. Filed June 8, 2018, and issued November 5, 2019. https://patents.google.com/patent/US10467239B2/en.

Karno, Marvin, and Donald A. Schwartz. *Community Mental Health: Reflections and Explorations*. Flushing, NY: Spectrum Publications, 1974.

Kauppi, Niilo. *French Intellectual Nobility: Institutional and Symbolic Transformations in the Post-Sartrian Era*. Albany, NY: SUNY Press, 1996.

Kauppi, Niilo. *The Making of an Avant-Garde: Tel Quel*. Berlin: Mouton de Gruyter, 1994.

Kay, Lily. "From Logical Neurons to Poetic Embodiments of Mind: Warren S. McCulloch's Project in Neuroscience." *Science in Context* 14, no. 15 (2001): 591–614.

Kay, Lily. *The Molecular Vision of Life: Caltech, the Rockefeller Foundation, and the Rise of the New Biology*. New York: Oxford University Press, 1993.

Kay, Lily. *Who Wrote the Book of Life? A History of the Genetic Code*. Stanford, CA: Stanford University Press, 2000.

Khilnani, Sunil. *Arguing Revolution: The Intellectual Left in Postwar France*. New Haven, CT: Yale University Press, 1993.

Kittler, Friedrich A. *Gramophone, Film, Typewriter*, trans. Geoffrey Winthrop-Young and Michael Wutz. Stanford, CA: Stanford University Press, 1999 [1986].

Kittler, Friedrich. "Flechsig/Schreber/Freud: An Informations Network of 1910" (1984). Translated by Laurence Rickels, Avital Ronell, David Levin, Adam Bresnick, Judith Ramme, and Friedrich Kittler. *Qui Parle* 2, no. 1 (Spring 1988): 1–17.

Kline, Nathan S., and Manfred Clynes. "Drugs, Space, and Cybernetics: Evolution to Cyborgs." In *Psychophysical Aspects of Space Flight*, edited by B. E. Flaherty, 345–71. New York: Columbia University Press, 1961.

Kline, Ronald R. *The Cybernetics Moment: Or Why We Call Our Age the Information Age*. Baltimore, MD: Johns Hopkins University Press, 2015.

Kline, Ronald R. "What Is Information Theory a Theory Of? Boundary Work among Scientists in the United States and Britain during the Cold War." In *The History and Heritage of Scientific and Technical Information Systems: Proceedings of the 2002 Conference, Chemical Heritage Foundation*, edited by W. Boyd Rayward and Mary Ellen Bowden, 15–28. Medford, NJ: Information Today, 2004.

Kline, Ronald R. "Where Are the Cyborgs in Cybernetics?" *Social Studies of Science* 39, no. 3 (June 2009): 331–62.

Kluckhohn, Clyde, S. L. Washburn, Gordon R. Willey, Eliot D. Chapple, Claude Lévi-Strauss, Ralph Beals, Stith Thompson, Robert H. Lowie, Ralph Linton, Margaret Mead, Harry Hoijer, Roman Jakobson, George P. Murdock, C. F. Voegelin, F. S. C. Northrop, J. O. Brew, J. Grahame Clark, and Hallam L. Movius Jr. "Pattern in Biology, Linguistics, and Culture" (roundtable discussion). In *An Appraisal of Anthropology Today*, edited by Sol Tax, Loren C. Eiseley, Irving Rouse, and Carl F. Voegelin, 299–321. Chicago: University of Chicago Press, 1953.

Koeneke, Rodney. *Empires of the Mind: I. A. Richards and Basic English in China, 1929–1979*. Stanford, CA: Stanford University Press, 2004.

Kohlbry, Marc. "Digital Index: Control Poetics in *Die Maschine*." *Symploke* 28, nos. 1–2 (2020): 83–99.

Kohlbry, Marc. "Feedback, Information, and the Fabrication of Spatialisme." *MLN* 135, no. 4 (2020): 859–87. https://doi.org/10.1353/mln.2020.0056.

Kristeva, Julia. "Semiotics: A Critical Science and/or a Critique of Science" (1969). In *The Kristeva Reader*, edited by Toril Moi, translated by Seán Hand, 74–89. New York: Columbia University Press, 1986.

Kuisel, Richard. *Seducing the French: The Dilemma of Americanization.* Berkeley: University of California Press, 1993.

Kurtz, Paul Michael. "The Philological Apparatus: Science, Text, and Nation in the Nineteenth Century." *Critical Inquiry* 47, no. 4 (Summer 2021): 747–76. https://doi.org/10.1086/714541.

Lacan, Jacques. "Discours de Rome" (1953). In *Autres écrits*, 133–64. Paris: Éditions de Seuil, 2001.

Lacan, Jacques. *The Ego in Freud's Theory and in the Technique of Psychoanalysis, 1954–1955.* Translated by Sylvan Tomaselli. New York: W. W. Norton, 1988.

Lacan, Jacques. "The Freudian Thing or the Meaning of the Return to Freud in Psychoanalysis" (1955 lecture, 1956 published). In *Écrits: The First Complete Edition in English*, translated by Bruce Fink, 334–63. New York: W. W. Norton, 2006.

Lacan, Jacques. "The Function and Field of Speech and Language in Psychoanalysis" (1953 lecture, 1956 published). In *Écrits: The First Complete Edition in English*, translated by Bruce Fink, 197–268. New York: W. W. Norton, 2006.

Lacan, Jacques. "Jeunesse de Gide ou la lettre et le désir" (1958). In *Écrits*, 739–64. Paris: Éditions du Seuil, 1966.

Lacan, Jacques. "The Mirror Stage as Formative of the I Function" (1949 version). In *Écrits: The First Complete Edition in English*, translated by Bruce Fink, 75–81. New York: W. W. Norton, 2006.

Lacan, Jacques. "Science and Truth" (1965 lecture, 1966 published). In *Écrits: The First Complete Edition in English*, translated by Bruce Fink, 726–45. New York: W. W. Norton, 2006.

Lacan, Jacques. "A Theoretical Introduction to the Functions of Psychoanalysis in Criminology" (1950 lecture, 1951 published). In *Écrits: The First Complete Edition in English*, translated by Bruce Fink, 102–22. New York: W. W. Norton, 2006.

Lacan, Jacques. "The Youth of Gide, or the Letter and Desire" (1958). In *Écrits: The First Complete Edition in English*, translated by Bruce Fink, 623–44. New York: W. W. Norton, 2006.

Lafontaine, Céline. *L'Empire cybernétique: Des machines à penser à la pensée machine.* Paris: Seuil, 2004.

Lasswell, Harold D., and Myres S. McDougal. "Legal Education and Public Policy: Professional Training in the Public Interest" (1943). In *The Analysis of Political Behaviour*, 21–131. London: Kegan Paul, Trench, Trubner, 1947.

Latour, Bruno. *Reassembling the Social: An Introduction to Actor-Network-Theory.* Oxford: Oxford University Press, 2005.

Latour, Bruno. "Why Has Critique Run Out of Steam? From Matters of Fact to Matters of Concern." *Critical Inquiry* 30, no. 2 (Winter 2004): 225–48.

Le Roux, Ronan. *Une histoire de la cybernétique en France (1948–1975).* Paris: Classiques Garnier, 2018.

Lee, Pamela M. *Think Tank Aesthetics: Midcentury Modernism, the Cold War, and the Neoliberal Present.* Cambridge, MA: MIT Press, 2020.

Lefebvre, Henri. "La notion de totalité dans les sciences sociales." *Cahiers Internationaux de Sociologie* 18 (January 1955): 55–77.

Lefebvre, Henri. "Marxisme et la théorie de l'information" (1958). In *Au-delà du structuralisme*, 51–75. Paris: Anthropos, 1971.

Lefort, Claude. "Qu'est-ce que la bureaucratie?" *Arguments* 4, no. 17 (1960): 64–81.

Lemov, Rebecca. *Database of Dreams: The Lost Quest to Catalog Humanity*. New Haven, CT: Yale University Press, 2015.

Lévi-Strauss, Claude. "Comparative Religions of Nonliterate Peoples" (1968). In *Structural Anthropology, Volume II*, 60–67. New York: Basic Books, 1976.

Lévi-Strauss, Claude. "Diogène couché" (1955). *Cités* 81, no. 1 (2020): 137–68.

Lévi-Strauss, Claude. *Introduction to the Work of Marcel Mauss*. Translated by Felicity Baker. 1950. London: Routledge, 1987.

Lévi-Strauss, Claude. "Language and the Analysis of Social Laws" (1951). In *Structural Anthropology*, 55–66. New York: Basic Books, 1963.

Lévi-Strauss, Claude. "Louis-Ferdinand Céline: Voyage au bout de la nuit" (1933). In *Claude Lévi-Strauss*, edited by Michel Izard, 23–24. Paris: Herne, 2004.

Lévi-Strauss, Claude. "The Mathematics of Man." *International Bulletin of Social Sciences* 6, no. 4 (1954): 581–90.

Lévi-Strauss, Claude. "The Place of Anthropology in the Social Sciences" (1954). In *Structural Anthropology*, 346–81. New York: Basic Books, 1976.

Lévi-Strauss, Claude. "Postscript to Chapter XV" (1958). In *Structural Anthropology*, 324–43. New York: Basic Books, 1976.

Lévi-Strauss, Claude. *Race and History*. Paris: UNESCO, 1952.

Lévi-Strauss, Claude. Review of "La grande épreuve des démocraties: Essai sur les principes démocratiques—leur nature—leur histoire—leur valeur philosophique," by Julien Benda. *Renaissance*, nos. 2–3 (1944): 324–28.

Lévi-Strauss, Claude. *The Savage Mind*. 1962. Chicago: University of Chicago Press, 1966.

Lévi-Strauss, Claude. "Le socialisme et la colonialisation." *L'étudiant socialiste*, October 1929, 7–8.

Lévi-Strauss, Claude. *Mythologiques*. Vols. 1–4. Paris: Plon, 1964–1971.

Lévi-Strauss, Claude. "Social Structure." In *Anthropology Today: An Encyclopedic Inventory*, edited by A. L. Kroeber, 524–53. Chicago: University of Chicago Press, 1953.

Lévi-Strauss, Claude. "Structural Analysis in Linguistics and Anthropology" (1945). In *Structural Anthropology*, 31–54. New York: Basic Books, 1976.

Lévi-Strauss, Claude. "The Structural Study of Myth." *Journal of American Folklore* 68, no. 270 (1955): 428–44. https://doi.org/10.2307/536768.

Lévi-Strauss, Claude. *Tristes Tropiques*. 1955. New York: Atheneum, 1975.

Lévi-Strauss, Claude, and Didier Eribon. *De près et de loin*. Paris: Odile Jacob, 1988.

Life. "Man Remade to Live in Space." July 11, 1960: 77–78.

Light, Jennifer S. "Discriminating Appraisals: Cartography, Computation, and Access to Federal Mortgage Insurance in the 1930s." *Technology and Culture* 52, no. 3 (July 2011): 485–522.

Light, Jennifer S. *From Warfare to Welfare: Defense Intellectuals and Urban Problems in Cold War America*. Baltimore, MD: Johns Hopkins University Press, 2003.

Lipset, David. *Gregory Bateson: The Legacy of a Scientist*. Englewood Cliffs, NJ: Prentice Hall, 1980.

Liu, Lydia H. *The Freudian Robot: Digital Media and the Future of the Unconscious*. Chicago: University of Chicago Press, 2010.

Lockett, William. "The Science of Fun and the War on Poverty." *Grey Room*, no. 74 (Winter 2019): 6–43.

Lombardo, Paul A. "Pioneer's Big Lie." *Albany Law Review* 66 (2002–3): 1125–44.

Loyer, Emmanuelle. *Lévi-Strauss: A Biography*. Translated by Ninon Vinsonneau and Jonathan Magidoff. Medford, MA: Polity, 2018.

Loyer, Emmanuelle. *Paris à New York: Intellectuels et artistes français en exil (1940–1947)*. Paris: Bernard Grasset, 2005.

Maisel, Albert. "Bedlam 1946." *Life*, May 6, 1946: 102–18.

Malcolm, Janet. "The One-Way Mirror." *New Yorker*, May 15, 1978: 39–114.

Manovich, Lev. *Cultural Analytics*. Cambridge, MA: MIT Press, 2020.

Marx, Leo. *The Machine in the Garden: Technology and the Pastoral Ideal in America*. New York: Oxford University Press, 1964.

Masani, Pesi R. *Norbert Wiener: 1894–1964*. Basel: Birkhäuser Verlag, 1990.

Matheson, Tamara Chaplin. "Embodying the Mind, Producing the Nation: Philosophy on French Television." *Journal of the History of Ideas* 67, no. 2 (2006): 315–42.

Mauss, Marcel. *The Gift: The Form and Reason for Exchange in Archaic Societies*. Translated by W. D. Halls. 1925. New York: Routledge, 1990.

Mauss, Marcel. "Les sciences sociales à Paris vues par Marcel Mauss" (1929). *Revue française de sociologie* 26, no. 2 (1985): 343–51.

Mauss, Marcel. "Techniques of the Body" (1934). *Economy and Society* 2, no. 1 (February 1973): 70–88.

Mazon, Brigitte. *Aux origines de l'école des hautes études en sciences sociales: Le rôle du mécénat américain (1920–1960)*. Paris: Les éditions du Cerf, 1988.

Mazon, Brigitte. "La fondation Rockefeller et les sciences sociales en France, 1925–1940." *Revue française de sociologie* 26, no. 2 (1985): 311–42. https://doi.org/10.2307/3321579.

McCormick, John P. *Carl Schmitt's Critique of Liberalism: Against Politics as Technology*. New York: Cambridge University Press, 1997.

McCulloch, Warren S. "A Recapitulation of the Theory, with a Forecast of Several Extensions." *Annals of the New York Academy of Sciences* 50, no. 4 (1948): 259–77. https://doi.org/10.1111/j.1749-6632.1948.tb39856.x.

McGinn, Lata K., and William C. Sanderson. "What Allows Cognitive Behavioral Therapy to Be Brief: Overview, Efficacy, and Crucial Factors Facilitating Brief Treatment." *Clinical Psychology: Science and Practice* 8, no. 1 (Spring 2001): 23–37.

McKelvey, Fenwick. "The *Other* Cambridge Analytics: Early 'Artificial Intelligence' in American Political Science." In *The Cultural Life of Machine Learning: An Incursion into Critical AI Studies*, edited by Jonathan Roberge and Michael Castelle, 117–42. London: Palgrave Macmillan, 2020. https://doi.org/10.1007/978-3-030-56286-1.

Mead, Margaret. *Coming of Age in Samoa: A Psychological Study of Primitive Youth for Western Civilization*. New York: William Morrow, 1928.

Mead, Margaret. "Cybernetics of Cybernetics." In *Purposive Systems: Proceedings of the First Annual Symposium of the American Society for Cybernetics*, edited by Heinz von Foerster, John D. White, Larry J. Peterson, and John K. Russell, 1–11. New York: Spartan Books, 1968.

Mead, Margaret. "From Intuition to Analysis in Communication Research." *Semiotica* 1, no. 1 (1969): 13–25. https://doi.org/10.1515/semi.1969.1.1.13.

Mead, Margaret. *Male and Female: A Study of the Sexes in a Changing World*. New York: William Morrow, 1967.

Mead, Margaret. *Soviet Attitudes toward Authority: An Interdisciplinary Approach to Problems of Soviet Character*. Santa Monica: The RAND Corporation, 1951.

Mead, Margaret. "The Study of Culture at a Distance." In Mead and Métraux, *The Study of Culture at a Distance*, 3–53.

Mead, Margaret. "The Training of the Cultural Anthropologist." *American Anthropologist* 54, no. 3 (1952): 343–46. https://doi.org/10.1525/aa.1952.54.3.02a00060.

Mead, Margaret, and Rhoda Métraux, eds. *The Study of Culture at a Distance*. Chicago: University of Chicago Press, 1953.

Mehlman, Jeffrey. *Emigré New York: French Intellectuals in Wartime Manhattan, 1940–1944*. Baltimore, MD: Johns Hopkins University Press, 2000.

Métraux, Rhoda. "The Study of Culture at a Distance: A Prototype." *American Anthropologist* 82, no. 2 (1980): 362–73.

Meynaud, Jean. "Qu'est-ce que la technocratie?" *Revue économique* 11, no. 4 (1960): 497–526. https://doi.org/10.3406/reco.1960.407420.

Meynaud, Jean. *Technocratie et politique*. Lausanne: Études de science politique, 1960.

Mills, Mara. "Deaf Jam: From Inscription to Reproduction to Information." *Social Text* 28, no. 102 (Spring 2010): 35–58. https://doi.org/10.1215/01642472-2009-059.

Mills, Mara. "Media and Prosthesis: The Vocoder, the Artificial Larynx, and the History of Signal Processing." *Qui Parle* 21, no. 1 (Fall–Winter 2012): 107–49.

Mindell, David A. *Between Human and Machine: Feedback, Control, and Computing before Cybernetics*. Baltimore, MD: Johns Hopkins University Press, 2002.

Mirowski, Philip. *Machine Dreams: Economics Becomes a Cyborg Science*. Cambridge: Cambridge University Press, 2002.

Mitchell, W. J. T. *Seeing Madness, Insanity, Media, and Visual Culture*. Berlin: Hatje Cantz Verlag, 2012.

Mitrani, Nora. "Bureaucratie et technicité." *Arguments* 4, no. 17 (1960): 28–30.

Moffat, Samuel. "Area Psychiatrists Study New Schizophrenia Theory." *Palo Alto Times*. May 7, 1958, second section.

Moretti, Franco. *Graphs, Maps, Trees: Abstract Models for a Literary History*. London: Verso, 2005.

Moretti, Franco. "'Operationalizing': Or, the Function of Measurement in Literary Theory." *New Left Review* 84 (December 2013): 103–19.

Mounin, Georges. "Communication, Linguistics and Information Theory" (1964). In *Semiotic Praxis: Studies in Pertinence and in the Means of Expression and Communication*, translated by Catherine Tihanyi, Maia Wise, and Bruce Wise, 33–49. New York: Plenum, 1985.

Müggenburg, Jan. "From Learning Machines to Learning Humans: How Cybernetic Machine Models Inspired Experimental Pedagogies." *History of Education* 50, no. 1 (2020): 112–33. https://doi.org/10.1080/0046760X.2020.1826054.

Muller, John P., and William J. Richardson. "Lacan's Seminar on 'The Purloined Letter': Overview." In *The Purloined Poe: Lacan, Derrida, and Psychoanalytic Reading*, edited by John P. Muller and William J. Richardson, 55–76. Baltimore, MD: Johns Hopkins University Press, 1988.

Museum of Modern Art. *Bali, Background for War: The Human Problem of Reoccupation*. Museum of Modern Art, New York, 1943. Accessed June 24, 2021. http://www.moma.org/calendar/exhibitions/3141.

Museum of Modern Art. "Museum of Modern Art Opens Exhibition of Bali, Background for War." Press release. Museum of Modern Art, New York, August 1943. http://www.moma.org/documents/moma_press-release_325409.pdf.

Needell, Alan A. "Project Troy and the Cold War Annexation of the Social Sciences." In *Universities and Empire: Money and Politics in the Social Sciences during the Cold War*, edited by Christopher Simpson, 3–38. New York: New Press, 1998.

Nemenyi, Daniel. "What Is an Internet? Norbert Wiener and the Society of Control." PhD diss., Kingston University, 2019.

Nessah, Rabia, Tarik Tazdaït, and Merhdad Vahabi. "The Game Is Afoot: The French Reaction to Game Theory in the 1950s," *History of Political Economy* 53, no. 2 (2021): 243–78.

Nobus, Dany. *Critique of Psychoanalytic Reason: Studies in Lacanian Theory and Practice*. New York: Routledge, 2022.

Nolan, Ginger. *The Neocolonialism of the Global Village*. Minneapolis: University of Minnesota Press, 2018.

Nordholt, Henk Schulte. "The making of traditional Bali: Colonial ethnography and bureaucratic reproduction." *History and Anthropology* 8, nos. 1–4: 89–127, https://doi.org/10.1080/02757206.1994.9960859.

Nye, David E. *American Technological Sublime*. Cambridge, MA: MIT Press, 1994.

Office of the Chief of Naval Operations. "Military Government Handbook: Kurile Islands." Washington: Office of the Chief of Naval Operations, November 1, 1943. http://archive.org/details/01150190R.nlm.nih.gov.

Olson, Richard. *Scientism and Technocracy in the Twentieth Century: The Legacy of Scientific Management*. Lanham, MD: Lexington Books, 2016.

O'Meara, Lucy. "List of Roland Barthes's Seminars and Lecture Courses at the École pratique des hautes études and the Collège de France, 1963–1980." In *Roland Barthes at the Collège de France*, edited by Lucy O'Meara, 205–6. Liverpool: Liverpool University Press, 2012.

O'Neill, John. "In Partial Praise of a Positivist." *Radical Philosophy* 74 (December 1995): 29–38.

Otis, Laura. *Networking: Communicating with Bodies and Machines in the Nineteenth Century*. Ann Arbor: University of Michigan Press, 2001.

Owens, Larry. "Mathematicians at War: Warren Weaver and the Applied Mathematics Panel, 1942–1945." In *The History of Modern Mathematics*, edited by David E. Rowe and John McCleary, 287–305. Boston: Academic Press, 1989.

Owens, Larry. "Vannevar Bush and the Differential Analyzer: The Text and Context of an Early Computer." *Technology and Culture* 27, no. 1 (January 1986): 63–95.

Paidipaty, Poornima. "'Tortoises All the Way Down': Geertz, Cybernetics and 'Culture' at the End of the Cold War." *Anthropological Theory* 20, no. 1 (2020): 97–129.

Pajon, Alexandre. *Claude Lévi-Strauss: Politique*. Descartes: Éditions Faustroll, 2007.

Pense, Arthur W., and Alfred M. Stanley. "The Institutional Care of Mentally Ill Children." *Psychiatric Services* 6, no. 9 (September 1955): 5–7. https://doi.org/10.1176/ps.6.9.5.

Peters, Benjamin. "Toward a Genealogy of a Cold War Communication Sciences: The Strange Loops of Leo and Norbert Wiener." *Russian Journal of Communication* 5, no. 1 (2013): 31–43. https://doi.org/10.1080/19409419.2013.775544.

Peters, John Durham. "Broadcasting and Schizophrenia." *Media, Culture and Society* 32, no. 1 (January 2010): 123–40.

Peters, John Durham. "'Memorable Equinox': John Lilly, Dolphin Vocals, and the Tape Medium." *boundary 2* 47, no. 4 (November 2020): 1–24. https://doi.org/10.1215/01903659-8677814.

Pias, Claus, ed. *Cybernetics: The Macy Conferences 1946–1953; The Complete Transactions*. Berlin: Diaphanes, 2016.

Pias, Claus. "The Age of Cybernetics." In Pias, *Cybernetics*, 11–26.

Piper, Andrew. *Can We Be Wrong? The Problem of Textual Evidence in a Time of Data Elements in Digital Literary Studies*. Cambridge: Cambridge University Press, 2020.

Pickering, Andy. "Cyborg History and the World War II Regime." *Perspectives on Science* 3, no. 1 (1995): 1–48.

Ponten, Frederic. "Collaborating with the Enemy: Wartime Analyses of Nazi Germany." PhD diss., Princeton University, 2017.

Porter, Theodore M. *Trust in Numbers: The Pursuit of Objectivity in Science and Public Life*. Princeton, NJ: Princeton University Press, 1996.

Pribilsky, Jason. "The Will to Enclose: Foucault's Archive in the Era of Cold War Big Data." *Le Foucauldien* 2, no. 1 (2016).

Price, David. *Anthropological Intelligence: The Deployment and Neglect of American Anthropology in the Second World War.* Durham: Duke University Press, 2008.

Price, David H. "Gregory Bateson and the oss: World War II and Bateson's Assessment of Applied Anthropology." *Human Organization* 57, no. 4 (1998): 379–84.

Queneau, Raymond. "Présentation de l'encyclopédie" (1955). In *Bords: Mathématiciens précurseurs encyclopédistes*. Paris: Hermann, 1963.

Quinan, Christine. "Technocrats and Tortured Bodies: Simone de Beauvoir's *Les Belles Images.*" *Women: A Cultural Review* 25, no. 3 (2014): 256–69. https://doi.org /10.1080/09574042.2014.981380.

Radin, Joanna. "'Digital Natives': How Medical and Indigenous Histories Matter for Big Data." *Osiris* 32, no. 1 (2017): 43–64. https://doi.org/10.1086/693853.

Raley, Rita. "Machine Translation and Global English." *Yale Journal of Criticism* 16, no. 2 (2003): 291–313. https://doi.org/10.1353/yale.2003.0022.

Rangil, Teresa Tomás. "The Politics of Neutrality: unesco's Social Science Department, 1946–1956." Rochester, NY: Duke University, Center for the History of Political Economy, 2011. https://doi.org/10.2139/ssrn.1840671.

Reidel, James. *Vanished Act: The Life and Art of Weldon Kees.* Lincoln: University of Nebraska Press, 2003.

Richards, I. A. "Variant Readings and Misreading" (1958 lecture). In *Style in Language*, edited by Thomas A. Sebeok, 241–52. Cambridge, MA: MIT Press, 1960.

Ricoeur, Paul. *Freud and Philosophy: An Essay on Interpretation.* Translated by Denis Savage. New Haven, CT: Yale University Press, 1970.

Riguet, Jacques. "Une analyse indolore: Propos recueillis par Deborah Gutermann-Jacquet." *Le diable probablement* 9 (2011): 98–102.

Robbe-Grillet, Alain. "On Several Obsolete Notions" (1957). In *For a New Novel: Essays on Fiction*, translated by Richard Howard, 25–47. Evanston, IL: Northwestern University Press, 1989.

Rockefeller Foundation. *The Rockefeller Foundation: A Review for 1936.* New York: Rockefeller Foundation, 1937.

Rockefeller Foundation. *The Rockefeller Foundation: Annual Report 1937.* New York: Rockefeller Foundation, 1938.

Rockefeller Foundation. *The Rockefeller Foundation Annual Report 1942.* New York: Rockefeller Foundation, 1943.

Rockefeller Foundation. *The Rockefeller Foundation: Annual Report 1948.* New York: Rockefeller Foundation, 1949.

Rockefeller Foundation. *The Rockefeller Foundation: A Review for 1950 and 1951.* New York: Rockefeller Foundation, 1951.

Rodowick, D. N. "An Elegy for Theory." *October*, no. 122 (Fall 2007): 91–109.

Rony, Fatimah Tobing. "The Photogenic Cannot Be Tamed: Margaret Mead and Gregory Bateson's *Trance and Dance in Bali.*" *Discourse* 28, no. 1 (Winter 2006): 5–27.

Ross, Dorothy. *The Origins of American Social Science*. Cambridge: Cambridge University Press, 1991.

Ross, Kristin. *Fast Cars, Clean Bodies: Decolonization and the Reordering of French Culture*. Cambridge, MA: MIT Press, 1995.

Rottenberg, Pierre. "Lectures de codes." In *Théorie d'ensemble*, 175–86. Paris: Éditions du Seuil, 1968.

Roudinesco, Elisabeth. *Jacques Lacan*. Translated by Barbara Bray. 1993. New York: Columbia University Press, 1997.

Rudy, Stephen. Introduction to Jakobson, *My Futurist Years*, vii–xxvi.

Rudy, Stephen. "Jakobson et Lévi-Strauss à New York (1941–1945), and Then Those Infamous Cats." In *Claude Lévi-Strauss*, edited by Michel Izard, 120–24. Paris: Herne, 2004.

Ruesch, Jurgen. "Values, Communication, and Culture: An Introduction." In *Communication: The Social Matrix of Psychiatry*, 3–20. New York: W. W. Norton, 1951.

Ruesch, Jurgen, and Weldon Kees. *Nonverbal Communication*. Berkeley: University of California Press, 1956.

Sack, Warren. "Une Machine à raconter des histoires: De Propp aux *software studies*," trans. Sophie Bargues Rollins, *Les Temps modernes* 676 (November–December 2013).

Sailer, Susan Shaw. "Universalizing Languages: *Finnegans Wake* Meets Basic English." *James Joyce Quarterly* 36, no. 4 (Summer 1999): 853–68.

Samoyault, Tiphaine. *Barthes: A Biography*. Translated by Andrew Brown. 2015. Cambridge: Polity, 2017.

Sarraute, Nathalie. *Portrait of a Man Unknown, a Novel*. Translated by Maria Jolas. 1948. New York: George Braziller, 1958.

Sartre, Jean-Paul. "Jean-Paul Sartre répond." *L'arc* 30 (October 1966): 87–96.

Saussure, Ferdinand de. *Course in General Linguistics*. Edited by Charles Bally and Albert Sechehaye. Translated by Wade Baskin. 1916. New York: Philosophical Library, 1959.

Schmitt, Carl. "The Age of Neutralizations and Depoliticizations" (1929). In *The Concept of the Political*, translated by M. Konzett and John P. McCormick, 80–96. Chicago: University of Chicago Press, 2007.

Schüttpelz, Erhard. "Quelle, Rauschen und Senke der Poesie: Roman Jakobsons Umschrift der shannonschen Kommunikation." In *Schnittstelle: Medien und Kommunikation*, edited by Georg Stanitzek and Wiljelm Voßkamp, 187–206. Cologne: DuMont, 2001.

Schwoch, James. *Global TV: New Media and the Cold War, 1946–69*. Urbana: University of Illinois Press, 2009.

Sconce, Jeffrey. *Haunted Media: Electronic Presence from Telegraphy to Television*. Durham, NC: Duke University Press, 2000.

Sconce, Jeffrey. *The Technical Delusion: Electronics, Power, Insanity*. Durham, NC: Duke University Press, 2019.

Sealander, Judith. *Private Wealth and Public Life: Foundation Philanthropy and the Reshaping of American Social Policy from the Progressive Era to the New Deal.* Baltimore, MD: Johns Hopkins University Press, 1997.

Segal, Jérôme. *Le zéro et le un: Histoire de la notion scientifique d'information au XXe siècle (Vol. I).* Paris: Éditions Matériologiques, 2013.

Serres, Michel. *The Parasite.* Translated by Lawrence R. Schehr. 1980. Minneapolis: University of Minnesota Press, 2007.

Shannon, Claude E. "Computers and Automation: Progress and Promise in the Twentieth Century" (1962 lecture). In *Claude Elwood Shannon: Collected Papers,* edited by N. J. A. Sloane and Aaron D. Wyner, 836–46. Piscataway, NJ: IEEE Press, 1993.

Shannon, Claude E. "Mathematical Theory of the Differential Analyzer." *Journal of Mathematics and Physics* 20, no. 4 (1941): 337–54. https://doi.org/10.1002/sapm1941201337.

Shannon, Claude E. "The Mathematical Theory of Communication" (1948). In *The Mathematical Theory of Communication,* 3–91. Urbana: University of Illinois Press, 1949.

Shannon, Claude E. "A Mind-Reading (?) Machine" (1953 memo). In *Claude Elwood Shannon: Collected Papers,* edited by N. J. A. Sloane and Aaron D. Wyner, 688–90. Piscataway, NJ: IEEE Press, 1993.

Shannon, Claude E. "The Potentialities of Computers" (1953 memo). In *Claude Elwood Shannon: Collected Papers,* edited by N. J. A. Sloane and Aaron D. Wyner, 691–94. Piscataway, NJ: IEEE Press, 1993.

Shannon, Claude E. "A Symbolic Analysis of Relay and Switching Circuits" (1938). In *Claude Elwood Shannon: Collected Papers,* edited by N. J. A. Sloane and Aaron D. Wyner, 471–96. Piscataway, NJ: IEEE Press, 1993.

Shapin, Steven, and Simon Schaffer. *Leviathan and the Air-Pump: Hobbes, Boyle, and the Experimental Life.* Princeton, NJ: Princeton University Press, 1985.

Shapiro, David A., Michael Barkham, William B. Stiles, Gillian E. Hardy, Anne Rees, Shirley Reynolds, and Mike Startup. "Time Is of the Essence: A Selective Review of the Fall and Rise of Brief Therapy Research." *Psychology and Psychotherapy: Theory, Research and Practice* 76, no. 3 (2003): 211–35.

Siedentrop, Larry. "Two Liberal Traditions." In *The Idea of Freedom: Essays in Honour of Isaiah Berlin,* edited by Alan Ryan, 153–74. Oxford: Oxford University Press, 1979.

Siegert, Bernhard. "Die Geburt der Literatur aus dem Rauschen der Kanäle: Zur Poetik der Phatischen Funktion." In *Electric Laokoon: Zeichen und Medien, von der Lochkarte zur Grammatologie,* edited by Michael Franz, Bernhard Siegert, and Robert Stockhammer, 5–41. Berlin: Akademie Verlag, 2007.

Soni, Jimmy, and Rob Goodman. *A Mind at Play: How Claude Shannon Invented the Information Age.* New York: Simon and Schuster, 2017.

Spielmann, Ellen, *Das Verschwinden Dina Lévi-Strauss' und der Transvestismus Mário de Andrades: Genealogische Rätsel in der Geschichte der Sozial- und Human-*

wissenschaften im modernen Brasilien. Berlin: Wissenschaftlicher Verlag, 2003.

Stanton, Alfred H., and Morris S. Schwartz. *The Mental Hospital: A Study of Institutional Participation in Psychiatric Illness and Treatment*. New York: Basic Books, 1954.

Stern, Steven. "Managed Care, Brief Therapy, and Therapeutic Integrity." *Psychotherapy: Theory, Research, Practice, Training* 30, no. 1 (1993): 162–75.

Stoczkowski, Wiktor. "Claude Lévi-Strauss and UNESCO." *Courrier de l'UNESCO* 5 (2008): 5–7.

Stoczkowski, Wiktor. "The 'Unconceivable Humankind' to Come: A Portrait of Lévi-Strauss as a Demographer." *Diogenes* 60, no. 2 (2013): 79–92. https://doi.org /10.1177/0392192114568268.

Sutton, Francis X. "The Ford Foundation's Transatlantic Role and Purposes, 1951–81." *Review (Fernand Braudel Center)* 24, no. 1 (2001): 77–104.

Tallman, Frank F. "Organization and Function of the Children's Group of Rockland State Hospital." *Psychiatric Quarterly* 12, no. 1 (March 1938): 175–88. https:// doi.org/10.1007/BF01561923.

Toman, Jindřich. *The Magic of a Common Language: Jakobson, Mathesius, Trubetzkoy, and the Prague Linguistic Circle*. Cambridge, MA: MIT Press, 1995.

Torracinta, Simon. "The Peripeteia of *The Gift*." Review of *Gift Exchange: The Transnational History of a Political Idea*, by Grégoire Mallard. *History of Anthropology Review* 44 (October 20, 2020). https://histanthro.org/reviews/peripeteia-of -the-gift/.

Touraine, Alain. "L'aliénation bureaucratique." *Arguments* 4, no. 17 (1960): 21–27.

Touraine, Alain. "Entreprise et bureaucratie." *Sociologie du travail* 1, no. 1 (1959): 58–71. https://doi.org/10.3406/sotra.1959.1003.

Touraine, Alain. "Le traitement de la société globale dans la sociologie américaine contemporaine." *Cahiers internationaux de sociologie* 16 (June 1954): 126–45.

Trumpener, Katie. "Critical Response I. Paratext and Genre System: A Response to Franco Moretti." *Critical Inquiry* 36, no. 1 (Autumn 2009): 159–71. https:// doi.org/10.1086/606126.

Turner, Fred. *From Counterculture to Cyberculture: Stewart Brand, the Whole Earth Network, and the Rise of Digital Utopianism*. Chicago: University of Chicago Press, 2006.

Turner, Fred. *The Democratic Surround: Multimedia and American Liberalism from World War II to the Psychedelic Sixties*. Chicago: University of Chicago Press, 2013.

US Department of State, "Foreign Affairs Research: Projects and Centers," Washington, D.C., November 15, 1960.

US Department of State. "Group Research Projects in Foreign Affairs and the Social Sciences." Washington, DC: Bureau of Intelligence and Research, September 10, 1958.

Underwood, Ted. *Distant Horizons: Digital Evidence and Literary Change*. Chicago: University of Chicago Press, 2019.

Valentini, Luísa. "Um laboratório de antropologia: O encontro entre Mário de Andrade, Dina Dreyfus e Claude Lévi-Strauss 1935–1938." PhD diss., Universidade de São Paulo, 2013. https://www.teses.usp.br/teses/disponiveis/8/8134/tde -06062011-132611/publico/2010_LuisaValentini.pdf.

Vardouli, Theodora. "Dissimilar at First Sight: Structural Abstraction and the Promises of Isomorphism in 1960s Architectural Theory." In *Instabilities and Potentialities: Notes on the Nature of Knowledge in Digital Architecture*, edited by Chandler Ahrens and Aaron Sprecher, 209–20. New York: Routledge, 2019.

Visser, Coert F. "The Origin of the Solution-Focused Approach." *International Journal of Solution-Focused Practices* 1, no. 1 (2013): 10–17.

von Foerster, Heinz, Margaret Mead, and Hans Lukas Teuber. "A Note by the Editors" (1952). In Pias, *Cybernetics*, 341–48.

von Neumann, John. *The Computer and the Brain*. 1958. New Haven, CT: Yale University Press, 2000.

Wang, Jackie. *Carceral Capitalism*. South Pasadena, CA: Semiotext(e), 2018.

Watter, Seth Barry. "Interaction Chronograph: The Administration of Equilibrium." *Grey Room*, no. 79 (Spring 2020): 40–77. https://doi.org/10.1162/grey_a_00292.

Watzlawick, Paul. "A Structured Family Interview." *Family Process* 5, no. 2 (September 1966): 256–71.

Weakland, John H. "One Thing Leads to Another." In *Rigor and Imagination: Essays from the Legacy of Gregory Bateson*, edited by Carol Wilder and John Weakland, 43–64. New York: Praeger, 1982.

Weakland, John H., Richard Fisch, Paul Watzlawick, and Arthur M. Bodin. "Brief Therapy: Focused Problem Resolution." *Family Process* 13, no. 2 (June 1974): 141–68.

Weaver, Warren. "The Average Reading Vocabulary: An Application of Bayes's Theorem." *American Mathematical Monthly* 27, no. 10 (1920): 347–54. https://doi.org/10 .2307/2972549.

Weaver, Warren. "Translation." In *Machine Translation of Languages: Fourteen Essays*, edited by W. N. Locke and A. D. Booth, 15–23. Cambridge, MA: MIT Press and Wiley, 1955.

Weaver, Warren. "Recent Contributions to the Mathematical Theory of Communication." In *The Mathematical Theory of Communication*, 94–117. Urbana: University of Illinois Press, 1949.

Weaver, Warren. "Statistical Freedom of the Will." *Reviews of Modern Physics* 20, no. 1 (1948): 31–34. https://doi.org/10.1103/RevModPhys.20.31.

Weber, Samuel. "The Calculable and the Incalculable: Hölderlin after Kittler." In *Media after Kittler*, edited by Eleni Ikoniadou and Scott Wilson, 35–50. London: Rowman and Littlefield, 2015.

Weber, Samuel. *Institution and Interpretation*. Minneapolis: University of Minnesota Press, 1987.

Wenner-Gren, Axel L. "Address of Welcome." In *An Appraisal of Anthropology Today*, edited by Sol Tax, Loren C. Eiseley, Irving Rouse, and Carl F. Voegelin, xiii–xiv. Chicago: University of Chicago Press, 1953.

Wiener, Anna. *Uncanny Valley: A Memoir*. London: Fourth Estate, 2020.

Wiener, Norbert. *Cybernetics, or Control and Communication in the Animal and the Machine*. Paris: Hermann and Cie, 1948.

Wiener, Norbert. *The Human Use of Human Beings: Cybernetics and Society*. 1950. Boston: Da Capo Press, 1988.

Wiener, Norbert. "What Is Information Theory?" IRE *Transactions on Information Theory* 2, no. 2 (June 1956): 48.

Weinstein, Deborah. *The Pathological Family: Postwar America and the Rise of Family Therapy*. Ithaca, NY: Cornell University Press, 2013.

Wilcken, Patrick. *Claude Lévi-Strauss: The Poet in the Laboratory*. New York: Penguin, 2010.

Wilder, Carol. "The Palo Alto Group: Difficulties and Directions of the Interactional View for Human Communication Research." *Communication Research* 5, no. 2 (Winter 1979): 171–86.

Wildes, Karl, and Nilo Lindgren. *A Century of Electrical Engineering and Computer Science at MIT, 1882–1982*. Cambridge, MA: MIT Press, 1985.

Williams, Raymond. *Television: Technology and Cultural Form*. 1974. New York: Routledge, 2003.

Wilson, Ara. "Visual Kinship." *History of Anthropology Newsletter* 42 (July 24, 2018). https://histanthro.org/clio/visual-kinship/.

Zolberg, Aristide R., and Agnès Callamard. "The École Libre at the New School, 1941–1946." *Social Research* 65, no. 4 (Winter 1998): 921–51.

Index

exchange: communications and, 120–21, 127, 134; culture as structured instance of, 63, 134; family interactions, 81; social media as, 175–79; gift-giving as, 113–22; international, 118; scientific knowledge as, 12, 27, 120; networks of, 27, 177–78; practices of, 63, 113, 141; structural theory of, 121–22, 127

Facebook: 10, 169, 176–79
Fahs, Charles, 120–21
family therapy: colonial research and, 66; cybernetics and, 16, 56–60; forerunners of, 38, 43–44, 46–47; game theory and, 79–84; Mead and Bateson's research on, 68–72; Palo Alto Group research on, 53–57, 76–84; postwar rise of, 73–84; social stability and, 69–70
"The Family as a Cultural Agent" (Frank), 43–44
fascism, 3, 28, 53, 157; family structure and, 69; structuralism as response to, 123
Federal Bureau of Investigation (FBI), 88, 117, 201n20
Feedback Mechanisms and Circular Causal Systems in Biology and the Social Sciences Meeting, 40–41. See also Macy Conferences
Fitzgerald, Margaret, 77
Ford Foundation, 22, 25, 118, 128, 129, 138, 140
Fosdick, Raymond, 21, 26
Foucault, Michel, 4, 134, 142, 145, 152, 156, 170, 183n16
Foundation Fund for Research in Psychiatry, 77
Frank, Lawrence, 42–45, 47, 49, 62, 67
Fremont-Smith, Frank, 41–43, 50
French-European American Foundation, 118. See also École des hautes études en sciences sociales (EHESS)
French theory, 3, 4; Barthes's contributions to, 158–64; communication research and, 153–58; cultural analytics and, 171–73, 175–76; cybernetics and, 136–40, 149–52, 167–68; information theory and, 5–6, 11; Lévi-Strauss and, 140–42; postwar emergence of, 133–36, 167–68; structuralism and,

147–49; technocracy and, 17–18, 110–14, 131, 142–47
Freud, Sigmund, 136, 142, 144–45, 149
Friedmann, Georges, 138, 155, 158
Fry, William F., 54, 77
functionalism, 121, 142; Barthes on 163–64
Fundamentals of Language (Jakobson & Halle), 98–99

Galison, Peter, 7, 29, 91
Galton, Francis, 37, 39
game theory, 44; Barthes and, 163–64; cybernetics and information theory in relation to, 11, 23–24, 30, 41, 52, 90, 98, 150, 166; definitions of, 8, 11–12; family research and, 78–84; language and, 98; Lévi-Strauss on, 120–28, 131, 137, 148; research funding for, 30, 41, 52; Shannon and Hagelbarger and, 146–48; structural linguistics and, 92, 98; of von Neumann, 131, 137
Gardin, Jean-Claude, 130
gender, 170, 171; literature and, 6; performance and, 9; shaping of science by, 165–66; structuralism and, 127, 165
genetics, 32, 36–37, 177–78. See also eugenics
"Generation That Squandered Its Poets, The" (Jakobson), 94–95
Genette, Gérard, 155
genocide, 1–2, 3, 20, 133, 176
German media theory, 8–9, 57. See also media theory
Gift: The Form and Reason for Exchange in Archaic Societies, The (Essai sur le don) (Mauss), 113–14, 122
Gift-giving: Lévi-Strauss and, 118, 121–22; Mauss on, 113–14; midcentury scientific correspondence as, 120; in social sciences, 123. See also exchange
Global South, 2, 116, 126, 150. See also Africa, Latin America, Indigenous peoples
Globalism: disorder resulting from, 108, 127; planetary-scale reform as, 1–2, 24, 27, 52, 62, 68, 123–27, 150; subjects of, 3; sciences in light of, 3–4, 9–10, 14, 27–28, 51–52, 130–32, 150, 151, 168; US and, 67, 91, 167

Lacan, Jacques: communications theory and, 4, 9, 12, 100, 120, 138, 152, 153, 155, 168, 173, 176; cultural analytics and, 176; French theory and, 17, 140–42, 165, 168; Guilbaud, 138; on human sciences, 19, 134; Lévi–Strauss and, 19, 130, 138, 141; operationalization and, 173; Rockefeller Foundation rejection, 96, 115; seminar of, 142–47; structural linguistics and, 92–93, 100; technocracy and human science and, 142–47

Langley Porter Neuropsychiatric Clinic, 71, 73–75, 77

language: codification of, 6, 21, 69–70, 158–62; communication theoretical jargon and, 11–14, 34, 50, 55, 119, 125, 136–37, 141, 153, 165–68, 174; cybernetics and, 98–106; Foucault on, 152; Lacan on, 143–47; poetics and, 102–106; politics and, 52, 91–92, 163–64; Russian, 90, 98–102; programming, 170; structuralist accounts of, 87, 93–98, 120–21, 177–78; science and, 21, 28–29; written language, 34, 59, 110

Lasswell, Harold A., 15, 19, 48

Latin America: US intelligence and, 117–18, 130; promoting science in, 49–50. *See also* Brazil

Laughlin, Harry, 37–38

Laura Spelman Rockefeller Memorial (LRSM), 26, 111

Lazarsfeld, Paul, 46, 48, 154, 188n26

Lazzarato, Maurizio, 23–24

Le Bras, Hervé, 130, 162

Lefebvre, Henri, 148–49, 155, 167

Lefort, Claude, 149

Le Lionnais, François, 141

Le Monde (newspaper), 137

Leontieff, Vassily, 105

Lévi–Strauss, Claude: administrative work of, 11–12, 122–28; communications theory and, 3–4, 9–12, 100, 107–8, 120–22, 147–49, 177; cultural analytics and, 170–79; cybernetics and, 9–10, 22, 85, 87–88, 120–22, 147–49, 154–55; datafication and the establishment of Laboratory of Social Anthropology, 109, 128–32; École libre des hautes études

en sciences sociales and, 114–20; human sciences and, 18; Lacan and, 144–45; political consulting and, 115–16, 156, 166; refugee in US, 8, 88, 90, 114–20; structuralism and, 17, 94, 92, 100, 128–32, 147–49, 153, 158, 169; technocracy and, 17–18, 143; UNESCO and, 122–28, 140–42

Lévi–Strauss, Dina (Dina Dreyfus), 107–8, 151

liberalism, 125–26, 171; aberrancy and, 59–60; cultural patterning and, 63–66; cybernetics and, 2–3, 6–10; democracy and, 6; family research and, 53–54, 59; Lévi–Strauss on, 125–26; neoliberalism and, 83; Mauss's rebuke of, 113–14, 122; robber baron philanthropy and, 27–28; technocracy and, 2, 7–10, 15–18, 24–27, 108, 111–14

Life magazine, 32–34, 35, *61*, 69–70

Light, Jennifer S., 7–8, 23–24

"Linguistic and Specular Communication: Genetic Models and Pathological Models" (Irigaray), 165

Linguistic Circle of New York, 85–90

Linguistic Circle of Prague, 85, 93–95, 106

Linguistic Circle of Moscow, 93, 95, 106

Linguistics, 18, 142, 173, 176; anti–Soviet applications of, 98–102; cybernetics and, 3, 10, 13, 41, 48, 95–98, 133, 141, 152; critical French receptions of, 156, 161, 165, 166; 144–45; poetics and, 102–106; political backdrops of, 91–95, 108–110, 125, 131–32; structural linguistics, 85–90, 108, 121, 134, 136, 151, 152; technocracy and, 17, 18, 23, 28

"Linguistics and Poetics" (Jakobson), 102–106

literary studies: Barthes' approach to, 6, 157–64; cybernetic themes in, 2, 6, 16, 22, 102–6; human sciences and, 18, 22; linguistics and, 88, 90, 93, 95, 98; poetics and, 102–6; posthumanism and, 10; statistical studies of, 174–75; technical influences in, 156–58

Locke, William, 100

logical positivism, 28–29, 188n28. *See also* Unity of Science movement

Loyer, Emmanuelle, 125

Riguet, Jacques, 140–41

Rimbaud, Arthur, 120

Rist, Charles, 112

Robbe–Grillet, Alain, 133

robber baron philanthropy: communications theory and, 3, 21–24, 52, 76–77; cultural analytics and, 173, 178; cybernetics and, 41–45, 58, 69; eugenics, genetics, and, 36–40; interwar computing and, 29–36; Palo Alto Group and, 76–84; postwar research programs and, 48–50, 125, 137–38; scientific philanthropy and, 25, 103–6; scientific research and, 24–29; technical reordering of sciences by, 24–29, 46–47, 102–3. See also Carnegie Institution, Carnegie Corporation, Ford Foundation, Guggenheim Foundation, Josiah Macy, Jr. Foundation, Macy Conferences, Rockefeller Foundation, Wenner–Gren Foundation

Rockefeller, John D., 25

Rockefeller, Nelson, 118

Rockefeller Analyzer, 30–31

Rockefeller Foundation: cybernetics research and, 42, 43, 45, 58; École libre des hautes études, 95–98 114–20, 127–28; integration of postwar research programs by, 48–50; interwar computing at MIT and, 39–36, 37, 40; Lévi–Strauss and, 120–21, 125, 127–28, 129, 143, 153; machine translation and, 50–53; Palo Alto Group and, 76–77; structural linguistics research and, 98–102, 166–67; technical reordering of science and, 21–29, 173; technocracy and, 3, 111–14; Unity of Science movement and, 29–30

Rockland State Hospital, 60, 61

Rodinson, Maxime, 131–32

Rodowick, David, 18

Rony, Fatimah Tobing, 63

Roosevelt, Franklin, 46, 90, 118

Rosenblueth, Arturo, 22, 41, 43, 58; Rockefeller support for, 49–52

Ross, Dorothy, 15, 18

Roudinesco, Elizabeth, 144

Ruesch, Jurgen, 71–76, 78

Russian Futurism, 90, 93, 94

Russian Research Center (Harvard University), 101–2

Sarrasine (Balzac), 6, 161, 163–64

Sarraute, Nathalie, 133

Sartre, Jean–Paul, 118, 149, 167–68

Saussure, Ferdinand de, 13, 85–87, 92–97, 99–100, 103, 110–11, 115, 136, 152

Saussure, Raymond de, 115–16

Savage Mind, The (Lévi–Strauss), 9, 150–55, 158

Schawinsky, Xanti, 67, *68*

schematic account of communication, 2, 51, 73, 102–6, 126, 152, 160–62, 165, 177. *See also* information theory, signal

Schizophrenia, 165; colonial research and, 62, 69–70, 100; family and, 36–38, 54–55, 75–76, 78–82; filmic studies of, 71, 174; postmodern ideas of, 55, 165

Schramm, Wilbur, 153–54

Schützenberger, M. P., 129, 141

Scientific American (magazine), 121

scientific philanthropy. *See* robber baron philanthropy

Sebeok, Thomas, 87, 103–104

SELECTO cards, 130. *See also* punched cards

semiology: Barthes and, 155–56, 160–64; postwar French thought and, 134, 136, 142, 153, 156, 164–66; cybernetics and, 2–4, 5, 10–11, 17; human sciences and, 18–19; digital technology and, 170–76; postwar technics and, 134, 136; structural linguistics and, 87–88, 90

Sender/receiver model of communication. *See* schematic account of communication

Serres, Michel, 134, 166

Shame of the States, The (Deutsch), 69

Shannon, Claude Elwood: Barthes appropriation concepts of, 153, 160–62; communications and information theory of, 12–13, 108, 129, 177, 148–49; cybernetics and, 41, 45, 77; Lacan on research of, 143, 146–47; genetics and eugenics research and, 36–40; interwar computing at MIT and, 31–36; mental health research and, 57–58; Lévi–Strauss and, 120; structural linguistics and, 86–90, 94, 98, 99, 102–105; wartime work of, 46; Weaver's reinterpretations of research of, 22–23, 28–29, 48, 51–52. *See also* schematic account of communication, information theory

"Shifters, Verbal Categories, and the Russian Verb" (Jakobson), 99

Shklovsky, Viktor Borisovich, 93

Signal, 5, 6, 14, 46, 54, 59, 74, 81–82, 103, 122, 146, 151, 161. *See also* schematic account of communication

Sixième section of Ecole pratique des hautes études (EPHE), 112, 118–19, 122, 138–39, 142, 144, 150, 158, 162. *See also* Ecole des hautes études en sciences sociales (EHESS); École pratique des hautes études (EPHE)

Social Science Research Council (SSRC), 62, 103

social sciences: cultural analytics and, 172–77; cybernetics, informatics and, 11, 12, 40–48, 57–60, 136; efforts to reform in France, 111–14, 119–28, 136–40, 140–42, 150, 152, 154, 155, 157–58, 167, 166; French–US comparisons of, 110–14, 136; technocratic reform and, 2–3, 7–8, 15–20, 21, 23, 26–27, 53, 57–9, 62, 148–49, 162; World War II and Cold War research in, 47, 66–69, *See also* anthropology, linguistics, human sciences

social services and, 7–8, 24, 57–58, 66–67

Sociology of Signs, Symbols, and Representations: Systems of Objects (clothing, food, housing) (seminar), 142

Sollers, Philippe, 162

Spivak, Gayatri, 174

structuralism: anthropology and, 107–109, 120–22, 128–32; Barthes and, 157–65; communication research and, 2, 13, 95–98; contests over cybernetics and, 147–49, 166–68; linguistics and, 85–90, 98–102; poetics and, 102–206; politics of, 91–95, 110–14, 116–20; psychoanalysis and, 142–47, 165–66; reconstruction in postwar France of, 138–42; technocracy and 17, 122–28

Study of Culture at a Distance project, 67

"Style in Language" Conference, 103–106

suburbs (US), 7, 8, 53 ,57, 68–72, 74–76, 77, 90

"Symbolic Analysis of Relay and Switching Circuits, A" (Shannon), 31–32, 36

symbolic order (Lacan), 144–47

System of Objects, The (Baudrillard), 166

S/Z S/Z(Barthes), 6, 161–64

Tati, Jacques, 133–34, *135*

technocracy: Barthes on, 157–60, 162–64; colonialism and, 110–14, 158–62; cultural analytics and, 170–71; interwar computing and, 29–36; cybernetics and, 7–11; defined, 14–15; eugenics, genetics and, 36–40; French social science, 108–114, 134, 136, 138, 148–50, 157, 167–68; human sciences and, 1–5, 55, 57, 58, 62–63, 69, 90; information theory and, 5–6; Lacan on, 142–47; Lévi-Strauss and, 108, 119, 125–28, 131–32; robber baron philanthropy and, 21–29, 40–45, 52, 115, 117; theory and, 5–6, 14–18

telephony. *See* American Telephone & Telegraphy (AT&T), Bell Telephone Laboratories, Materiel téléphonique, le

Tel Quel (journal), 149, 153

Teuber, Hans Lukas, 50

theory: cybernetics and the rise of, 1–4; definitions and function of, 5–8, 10, 16; Baudrillard's critique of, 166–67; French theory, 134–36, 168, 170–72, 175; machine modeling of, 27, 31–2, 148–49; technocracy and, 14–18, 110–14, 119–20, 131, 142–47. *See also* communication theory, critical theory, French theory, game theory, information theory, media theory

Tinguely, Jean, 133–34, *135*

Todorov, Tzvetan, 155

Touraine, Alain, 148–49

"Toward the Logical Description of Languages in Their Phonemic Aspect" (Cherry, Halle, and Jakobson), 99

Tristes Tropiques (Lévi-Strauss), 116–17, 126, 151, 164

Trubetzkoy, Nikolai, 93–95

Trumpener, Katie, 174